Diseases and Disorders of Ornamental Palms

A. R. Chase and T. K. Broschat, editors

APS PRESS
The American Phytopathological Society
St. Paul, Minnesota, USA

Financial Sponsors

South Florida Chapter of the Palm Society, Inc.
University of California Tree Management Seminar
Department of Plant Pathology, University of California, Riverside
Department of Plant Pathology, University of Florida

Front cover: *Phoenix reclinata* (foreground) and *Washingtonia robusta* (background). Photo courtesy of T. K. Broschat.

Back cover: Specimen of *Euterpe edulis*. Photo courtesy of T. K. Broschat.

This book has been reproduced directly from computer-generated copy submitted in final form to APS Press by the editors of the volume. No editing or proofreading has been done by the Press.

Reference in this publication to a trademark, proprietary product, or company name by personnel of the U.S. Department of Agriculture or anyone else is intended for explicit description only and does not imply approval or recommendation to the exclusion of others that may be suitable.

Library of Congress Catalog Card Number: 91-70724
International Standard Book Number: 0-89054-119-1

© 1991 by The American Phytopathological Society

All rights reserved.
No portion of this book may be reproduced in any form, including photocopy, microfilm, information storage and retrieval system, computer database, or software, or by any means, including electronic or mechanical, without written permission from the publisher.

Copyright is not claimed in any portion of this work written by U.S. government employees as a part of their official duties.

Printed in the United States of America

The American Phytopathological Society
3340 Pilot Knob Road
St. Paul, Minnesota 55121, USA

Contributors

M. Aragaki, Department of Plant Pathology, University of Hawaii, Honolulu
T. K. Broschat, Ft. Lauderdale Research and Education Center, University of Florida
A. R. Chase, Central Florida Research and Education Center, University of Florida
H. D. Ohr, Department of Plant Pathology, University of California, Riverside
G. W. Simone, Department of Plant Pathology, University of Florida
J. Uchida, Department of Plant Pathology, University of Hawaii, Honolulu

TABLE OF CONTENTS

vii	Preface

DISEASES

1	Algal leaf spot - G. W. Simone
2	Annellophora leaf spot - G. W. Simone
2	Anthracnose (Colletotrichum leaf spot) - A. R. Chase
3	Bacterial bud rot - G. W. Simone
5	Bipolaris, Exserohilum, and Phaeotrichoconis leaf spots - M. Aragaki, A. R. Chase and J. Uchida
7	Calonectria leaf spot (Cylindrocladium leaf spot) - M. Aragaki and J. Uchida
8	Catacauma leaf spot (Tar spot) - G. W. Simone
10	Damping-off - A. R. Chase
10	Diamond scale - H. D. Ohr
11	Fusarium wilt - H. D. Ohr
12	Ganoderma butt rot (Basal stem rot) - G. W. Simone
14	Graphiola leaf spot (False smut) - G. W. Simone
16	Lethal yellowing disease (Awka disease, Cape St. Paul wilt, Kaincope disease, Kribi disease, and Pudricion del Cogollo) - G. W. Simone
20	Mosaic - G. W. Simone
20	Pestalotiopsis leaf spot - G. W. Simone
21	Phytophthora diseases - M. Aragaki, G. W. Simone and J. Uchida
24	Pink rot (Gliocladium blight) - H. D. Ohr
26	Pseudocercospora and Cercospora leaf spots - M. Aragaki and J. Uchida
27	Pseudomonas blight - A. R. Chase
28	Rachis blight - H. D. Ohr
29	Sclerotinia blight - A. R. Chase
29	Stigmina leaf spot (Exosporium leaf spot) - G. W. Simone
30	Thielaviopsis bud rot (Stem bleeding, Bitten leaf, Black scorch, Dry basal rot, Heart rot) - G. W. Simone

PHYSIOLOGICAL DISORDERS
- T. K. Broschat

33	Albinism
33	Boron deficiency
34	Boron toxicity
34	Calcium deficiency
34	Chlorine deficiency
35	Cold injury
36	Copper deficiency
36	Copper toxicity
37	Excessive water uptake
37	Fluoride toxicity
38	Foliar salt injury
38	Hapaxanthic flowering
38	Herbicide toxicity
40	High soil soluble salts
40	Iron deficiency
41	Lightning injury
41	Magnesium deficiency
42	Manganese deficiency
43	Nitrogen deficiency
43	Phosphorus deficiency

44	Potassium deficiency
46	Powerline decline
46	Root suffocation
47	Sulfur deficiency
47	Sunburn
48	Water stress
48	Zinc deficiency

APPENDICES

49	Appendix 1. Chu medium No. 10 for culturing *Cephaleuros*
49	Appendix 2. Palm taxa reported as hosts of *Bipolaris* spp.
49	Appendix 3. Palm taxa reported as hosts of *Exserohilum rostratum* or *Phaeotrichoconis crotalariae*
50	Appendix 4. Palm taxa known to be hosts of *Ganoderma zonatum, Graphiola phoenicis,* or *Pestalotiopsis palmara*
51	Appendix 5. Susceptibility of palm taxa to Lethal Yellowing Disease
52	Appendix 6. Antibiotic injection methods for control of Lethal Yellowing Disease
52	Appendix 7. Palm taxa reported as hosts of *Phytophthora palmivora*
52	Appendix 8. Some amendments to corn meal agar used to isolate *Phytophthora* spp.
53	Appendix 9. Palm taxa known to be hosts of *Stigmina palmivora*
53	Appendix 10. Critical concentrations of 13 elements in *Chamaedorea elegans, C. erumpens* and *Chrysalidocarpus lutescens*
54	Appendix 11. Critical concentrations of 13 elements in *Howea forsterana* and *Rhapis excelsa*
54	Appendix 12. Critical concentrations of 13 elements in *Elaeis guineensis*
55	Appendix 13. Common names of some palm taxa

PREFACE

Diseases and Disorders of Ornamental Palms was first conceived following a Symposium held at the 1988 annual meeting of the American Phytopathological Society. Each of the contributing authors presented talks featuring the best color slides available on palm diseases and disorders. We wanted to make these illustrations available to landscapers, producers and scientists to diagnose palm problems. We included micrographs of the pathogens when possible to further aid the diagnostician. We have tried to include all known diseases and disorders of ornamental palms including a few whose cause is poorly understood. Some portions of information which we gathered could not be included in this book due to space constraints. A list of literature citations as well as a listing of botanical gardens with palm collections worldwide can be obtained from A. R. Chase, CFREC-Apopka, 2807 Binion Rd., Apopka, FL 32703.

The contributors are directly responsible for the timeliness of this publication since much of the information contained herein has not been published elsewhere. Many thanks for their patience and continued good humor. I am especially grateful to Howard Ohr and Gary Simone for financial efforts which made publication of this book possible. We also owe thanks to the three Universities who supported the efforts of the contributors in many ways.

Finally, special thanks are extended to the following people who also contributed to the quality of this book.

Gary Chastagner, Washington State University, West Washington Research-Extension Center - Puyallup
Nancy Cochrane, University of Florida, Central Florida Research and Education Center - Apopka
Jim Downer, University of California, Cooperative Extension Service, Ventura County
Patrick Fenn, University of Arkansas, Fayetteville
Nigel Harrison, University of Florida, Ft. Lauderdale Research and Education Center
Alan Meerow, University of Florida, Ft. Lauderdale Research and Education Center
APS Press Staff, American Phytopathological Society, St. Paul, Minnesota

DISEASES

ALGAL LEAF SPOT
(Red Rust or Algal Rust)

Symptomology: Young lesions first appear as yellow pinpoint spots primarily on upper leaf surfaces. These lesions expand into crusty appearing, gray-green patches that will assume an orange cast during reproductive periods of the pathogen. On *Elaeis*, high incidence of this disease may cause up to 20% premature senescence of entire leaves. The pathogen may invade pinnae and the rachis of palms and invasion of stem and fruit tissue is known on some crops.

Causal Organism: *Cephaleuros virescens* (a green alga) is the most important of the plant parasitic algae. Initial infection occurs toward the end of humid, rainy seasons when rain disseminates the biflagellate zoospores from sporangiophores found in old infections. Algal thalli penetrate leaves by mechanical force and grow chiefly along the horizontal plane between the leaf cuticle and the epidermal cell layers. Vegetative growth continues slowly for approximately 8-9 months until rains and/or high humidity trigger a reproductive phase. Reproductive structures develop the pigment haematochrome that is responsible for development of the orange color on the algal spots. Lesions remain fertile throughout the leaf age and probably for several months after leaf abscission.

Occurrence: Algal leaf spot has been reported worldwide between latitudes 32°N and 32°S on many hosts. Algal spot on palms has been reported from Australia, Brazil, China, Costa Rica, Honduras, India, Japan, Nicaragua, Nigeria, Puerto Rico, United States (Florida and Hawaii), and the West Indies.

Species Affected: *Cephaleuros* is reported on *Bactris gasipaes*, *Butia capitata*, *Caryota* spp., *Cocos nucifera*, *Elaeis guineensis*, *Phoenix dactylifera*, *Sabal palmetto*, and *Trachycarpus fortunei*. In addition, this algal species has an extremely wide host range that includes such economically important species as avocado, cacao, cashew nut, citrus, coffee, guava, litchi, mango, pepper, rubber, and tea.

Diagnostic Techniques: Microscopic examination of free-hand sections of algal spots will reveal the subcuticular cells of the thallus. The pigmented aerial trichomes or sporangiophores with sporangia may also be present. The alga can be cultured in a medium developed by Chu (No. 10) given in Appendix 1. Light exposure is assumed but is undefined in the literature. Growth in culture from either zoospores or pieces of thallus is slow. Secondary fungi such as *Colletotrichum* sp. may invade empty algal reproductive structures on some host species and thereby confuse isolation results.

Prevention and Treatment: Incidence of algal spot on some non-palm hosts seems to be correlated with either low plant vigor or planting sites with high humidity and poor air circulation. Selective pruning and thinning of plant canopies can increase air circulation and reduce incidence. Algal spot on some food crops has been successfully controlled by applications of copper fungicides timed for the end of the rainy season to prevent zoospore infection.

Algal leaf spot on *Podocarpus nagi*.

Closeup of *Cephaleuros* on *P. nagi*.

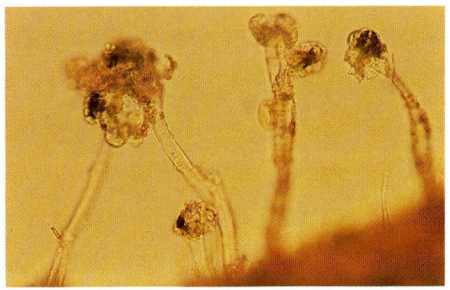

Cephaleuros virescens closeup.
(Courtesy R. Cullen)

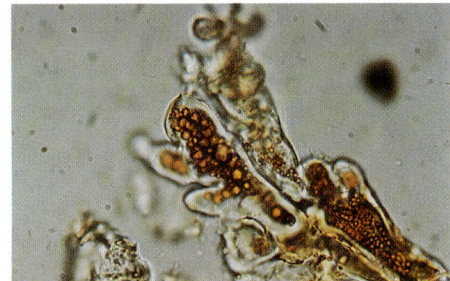

Cephaleuros virescens closeup.
(Courtesy Fl. Div. Plant Ind.)

Nonparasitic green alga on leaves.

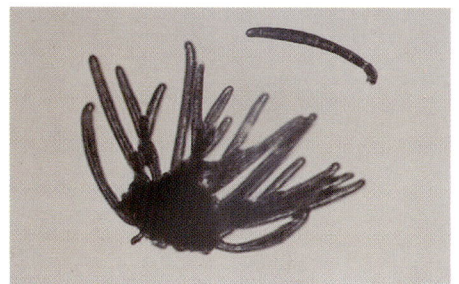

Annellophora leaf spot on *Phoenix canariensis*.

Micrograph of *Annellophora phoenicis*.

Anthracnose on *Washingtonia* sp.

Anthracnose on *Caryotis mitis*.

Anthracnose on *Paurotis* sp.

ANNELLOPHORA LEAF SPOT

Symptomology: Leaf spots are circular, brown in color and may reach a diameter of 1 cm. Lesions often have dark brown to black centers with a narrow yellow halo. When infections are numerous, lesions will merge causing distortion and/or death of leaf pinnae. Leaf spots resemble those caused by such other fungi as *Cercospora* and *Stigmina* spp.

Causal Organism: *Annellophora phoenicis* is a dematiaceous hyphomycete. Mycelia are nearly colorless to pale brown. Conidia are produced from either solitary or grouped conidiophores which can be produced terminally or laterally from mycelial strands. Conidiophores are brown, swollen at the base, with up to 7 cells, measuring 30-70 µm long and 3-5 µm thick. Conidia are borne singly from the tip of the conidiophore. Conidia are pale brown to brown in color, straight to slightly curved, with 8-14 cells, and measure 50-70 µm in length x 11-13 µm in width (base) to 2-3 µm in width (apex).

Occurrence: This disease is reported from Sierra Leone, Malaysia, New Guinea, and the United States (Texas).

Species Affected: Annellophora leaf spot has been reported on both *Phoenix canariensis* and *P. reclinata*.

Diagnostic Techniques: The pathogen will sporulate from either leaf surface. Infected leaf tissue can be surface disinfested in either a 0.5% sodium hypochlorite solution or in a 70% ethanol plus 0.5% sodium hypochlorite solution for 3 min. Abundant sporulation along the lesion margins should be evident within 5 days. On potato-dextrose agar colonies are slow growing (like *Cercospora*) are dark brown and have an entire margin.

Prevention and Treatment: Annellophora is known to colonize wounded as well as healthy tissue. Avoid foliar damage from abrasion, chemical burn, or other foliar diseases. Reduction of or avoidance of overhead irrigation will minimize periods of leaf wetness favorable for infection. Removal of infected tissue will reduce the reservoir of inoculum and slow disease spread. Although, no data are available, fungicides such as chlorothalonil, iprodione, maneb, and mancozeb should be effective in reducing this foliar disease. Select flowable formulations wherever possible or add a surfactant to wettable powder formulations to improve coverage and adherence of fungicide to leaves.

ANTHRACNOSE
(Colletotrichum Leaf Spot)

Symptomology: Symptoms vary somewhat from host to host and can be small, water-soaked speckles or large necrotic and chlorotic lesions which are circular to irregular in shape. They are usually tan to black with a bright yellow chlorotic halo and generally coalesce as symptoms progress. On some hosts the lesions are angular since they are confined to the area between leaf veins. On other hosts a frogeye spot may develop. Bipolaris leaf spot of palms is frequently confused with Colletotrichum leaf spot due to the similarity in symptoms. Young plants are especially susceptible to the pathogen perhaps because of the presence of immature foliage or the extremely humid environment created in the seedling trays. Damaged seedlings may be lost quickly if control measures are not taken.

Causal Organism: *Colletotrichum gloeosporioides* is most frequently associated with these leaf spots. Isolates are somewhat variable and usually handled as a group. Conidia are 9-24 x 4-6 µm, straight and obtuse at the apex. They readily form acervuli on host tissue which are clearly seen as tiny black specks, sometimes formed in concentric rings in the necrotic lesions. Acervuli are not always present in new lesions. The characteristic setae may or may not be formed on the leaf tissue or *in vitro*. Isolation is very easy on standard nutrient medium such as potato-dextrose agar or vegetable juice agar. Attributing the cause of a leaf spot to *C. gloeosporioides* should proceed with caution since this fungus is an excellent saprophyte and is commonly associated with necrotic tissue of many diseases and disorders.

Occurrence and Species Affected: Very little work has been reported regarding anthracnose of palms although *Colletotrichum* has been recovered from most species grown in the majority of palm-producing areas in the world. A report of Colletotrichum leaf spot has been made from Australia on *Phoenix roebelenii*.

Diagnostic Techniques: Anthracnose fungi are very easy to recover from symptomatic leaves using standard surface-disinfestation and plating techniques.

Prevention and Treatment: Any method which keep foliage dry reduces the potential for infection and spread of the pathogen. Do not use overhead irrigation or expose plants to rainfall if possible. Preventive applications of mancozeb fungicides can greatly reduce incidence of the disease. Improve rapid drying of the foliage by spacing plants adequately and rogue severely infested flats.

BACTERIAL BUD ROT

Symptomology: The initial symptom is a dark-brown, wet rot on the lower portion of the unopened spear. Disease incidence often coincides with recent cold damage, or a drastic change in moisture availability. The rot spreads into the pinnae as the spear opens. Rain splash is believed to spread the infection to other fronds. The bacterium continues to move down the rachis of the spear leaf until the spear collapses and hangs downward from the crown. If the rot continues down into the bud, the bud develops a smelly soft decay and the palm dies. In many cases, however, the infection is walled off by a callus layer formed above the bud, and bud recovery begins. The recovery period is characterized by the emergence of numerous 'little leaves' that grow more normal in appearance with each emergence.

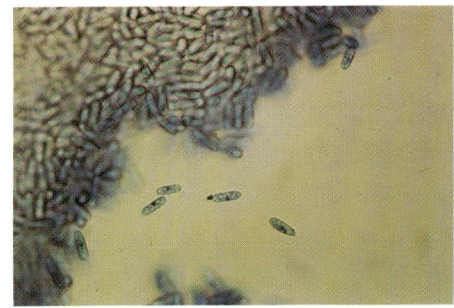

Conidia of *Colletotrichum* sp. (Courtesy D. Brunk)

Acervulus of *Colletotrichum* sp. (Courtesy D. Brunk)

Bacterial bud rot of *Neodypsis* sp.

Bacterial bud rot of *Cocos nucifera*. (Courtesy N. Harrison)

Rotted spear from *Neodypsis* sp. from bacterial bud rot.

Palm recovering from bacterial bud rot.

Bipolaris incurvata on *Howea forsterana*.

Leaf shredding on *Chrysalidocarpus lutescens* from *Bipolaris setariae*.

Causal Organism: Identity of the pathogen of bacterial rot is unclear. Seven genera have been implicated in the bud rot disease in Asia and the Pacific region including *Bacillus, Enterobacter, Escherichia, Leuconosta, Pseudomonas, Serratia,* and *Streptococcus.* In addition, *Erwinia carotovora* has been implicated in this disease in Cuba, *Erwinia* sp. in Florida, and Columbia and *E. lathyri* in Zaire. Revised taxonomy equates *E. lathyri* with *E. herbicola* and its presence in palm buds is usually viewed as saprophytic. The causal role of one or more of these bacteria remains to be demonstrated.

Occurrence: Reports of this disease in the older literature are now viewed somewhat doubtfully. Worldwide distribution is, therefore, also in question. Reports involving a suspected bacterial bud rot of palm exist from Columbia, Cuba, Ecuador, Nigeria, Republic of the Congo, southeast Asia, the United States (Florida), and Zaire.

Species Affected: Because of the confusion in the literature concerning the role of a bacterium in bud rot, a host range is difficult to assess. Reports of a bacterial soft rot of the bud exist for *Elaeis guineensis, Phoenix canariensis,* and *Roystonea* spp.

Diagnostic Techniques: Since the role of a specific pathogen is unknown, symptomatic tissue should be broadly processed to rule out a number of other possible pathogens. Tissue from the edge of the advancing bud decay should be sampled to avoid secondary microbes. This tissue should be streaked across nutrient agar and also across a pectin-containing medium or a cut potato tuber to evaluate pectate degradation. In addition, tissue pieces should be suitably surface disinfested and plated onto potato-dextrose agar and a selective medium for *Phytophthora* spp. The involvement of either *Chalara (Thielaviopsis) paradoxa* or *Phytophthora palmivora* should be determined. Subsequent recoveries of pectolytic bacterial colonies should be fully characterized to known *Erwinia* spp. Incidence of *Erwinia* spp. from symptomatic tissue should be correlated with recent environmental conditions since a number of bacteria are likely to be secondary invaders of weather-stressed palms.

Prevention and Treatment: Incidence of bacterial bud rot has been correlated with environmental conditions such as the transition from the dry season to the wet season. Treatment measures involve both sanitation and copper-based fungicides. When the rot is found on the exposed spear, the tissue can be excised before the disease spreads into the bud tissue. Buds of oil palms have been treated with sprays of either copper-Bordeaux mixtures or copper sulfate in South America and Africa. In addition, a copper-based paste has been painted onto the spear both as a deterrent and a curative practice. Bud drenches of a copper-based fungicide are commonly used to protect palm buds in the southeastern U.S. after cold damage. Since the role of bacteria in bud decay is still unknown, the benefit of drenches is also unknown.

BIPOLARIS, EXSEROHILUM, AND PHAEOTRICHOCONIS LEAF SPOTS

Symptomology: Leaf spots caused by the organisms in this category are generally similar to each other and are frequently sufficiently severe to drastically reduce marketability of many ornamental palms. Leaf spots begin as tiny (0.5 mm) water-soaked spots which may be chlorotic to greenish-brown, expand into circular to elliptical lesions, (2 to 10 mm or more in length), becoming brown, reddish-brown, or dark-brown to black in color. Lesions may be surrounded by a chlorotic, circular or spindle-shaped halo, conspicuous on some palms (e.g. *Howea*), but less distinct on others. Depending on host species, lesions may be slightly sunken or "eye spot" symptoms may develop as lesions expand. In severe cases, young leaves are blighted by coalescing lesions and fronds appear shredded. Severely blighted seedlings may be killed.

In spite of the general similarity in symptoms, there are some differences due to host reaction and pathogen involved. Lesions are commonly dark brown to black on *Chamaedorea*, whereas reddish-brown lesions are common on *Chrysalidocarpus*. Tissue yellowing and chlorosis, mentioned above, are more frequent in some palm genera than others.

Causal Organisms: *Exserohilum rostratum*, several species of *Bipolaris*, and *Phaeotrichoconis crotalariae* are known to cause leaf spots on palms. The first two genera have been referred to as *Helminthosporium*, then *Drechslera*. *Drechslera* has been separated into *Drechslera*, *Bipolaris*, and *Exserohilum* with the known palm pathogens falling into the late two genera.

Accurate identification of these fungi depends on environmental controls during culture growth and sporulation. *Exserohilum*, *Bipolaris*, and many other fungal species produce one spore type in the dark, another type under illumination, and mixed or intermediate types under diurnal light changes. Copious conidial production can be obtained by growing these fungi on vegetable juice agar (10% Campbell's V-8 juice, 0.2% $CaCO_3$ plus 1.7% agar), under approximately 2,500 lux continuous illumination from fluorescent lamps at about 24°C. Identification can be made in 5-7 days.

Exserohilum rostratum produces long (ca .200 μm), obclavate, rostrate, light olivaceous to gray-black conidia under continuous light conditions. Spore diameter averages 13 μm at the widest point. Dark-formed conidia average 100 μm or less in length, are elongate-fusoid with rounded ends and darker in color. Both spore forms are pseudoseptate, and end cells are frequently lighter in color and delimited by dark septa. The protuberant conidial hilum distinguishes this genus. Colonies on vegetable juice agar are initially dark gray and felty, turning brownish black with age.

In light, *Bipolaris setariae* produces conidia which are mostly curved, fusoid to elongate-ellipsoid, with rounded apex and base, mostly 50-70 by 10-13 μm with 5-9 pseudosepta, and dark olivaceous-brown at maturity. Dark formed conidia are elongate-ellipsoid, 40-60 x 10-14 μm, with 4-9 pseudosepta, becoming grayish olive-brown at maturity.

Exserohilum rostratum on *Chrysaidocarpus lutescens*.

Phaeotrichoconis crotalariae on *Livistona chinensis*.

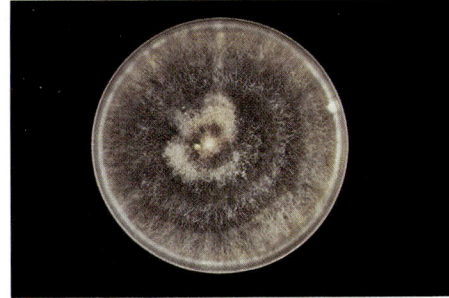

PDA culture of *Bipolaris setariae*.

PDA culture of *Exserohilum rostratum*.

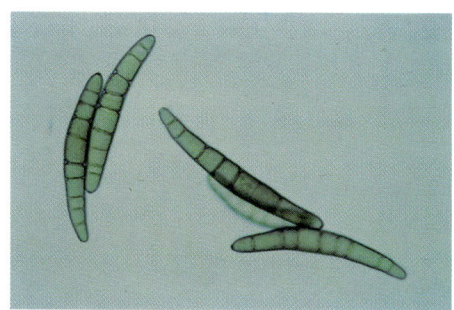

Conidia of *Bipolaris setariae* produced under light.

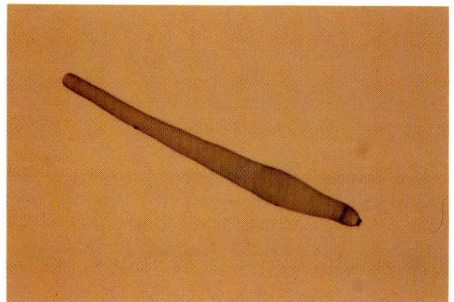

Conidium of *Exserohilum rostratum*.

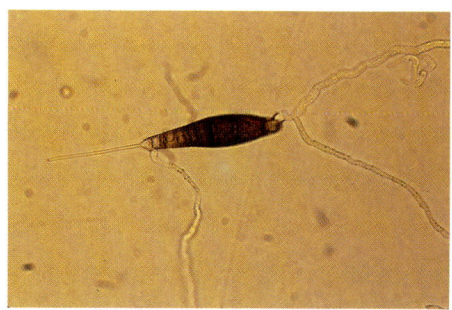

Conidium of *Phaeotrichoconis crotalariae*.

Leaf spot of *Howea forsterana* caused by *Calonectria theae*.

Under illumination at 24°C, *B. incurvata* produces well developed conidiophores but few spores. By transferring colonies to 20°C, curved, fusoid conidia tapering to a rounded apex, averaging 130 by 17 µm with 6-14 pseudosepta, and gray to dark olive brown in color, will be formed.

Bipolaris cynodontis produces slightly curved, fusoid conidia with tapered, blunt ends, averaging 50 by 13 µm, with 7-8 pseudosepta, pale to mid-golden brown with thin smooth walls. Colonies on vegetable juice agar medium are initially white with abundant aerial mycelium and later turn various shades of gray.

Phaeotrichoconis crotalariae is characterized by unbranched conidiophores which are straight or flexous, brown and smooth. Conidia are 50-85 x 15-22 µm at their broadest point, solitary, dry, obclavate, rostrate, and transversely septate. They have a golden brown body with a large dark brown scar. Colonies on vegetable juice agar are velvety, black, and have an irregular margin.

Occurrence and Species Affected: *Bipolaris* spp. are common on Arecaceae worldwide. *Bipolaris incurvata* has been reported causing leaf spots on *Cocos nucifera* in Fiji, French Polynesia, Malaysia, Philippines, Vietnam, Thailand, Jamaica, Seychelles, Hawaii, and many other places. On ornamental palms, *B. incurvata*, *B. setariae*, *B. cynodontis*, *B. australiensis*, *B. maydis* (anamorph of *Cochliobolus heterostrophus*), *B. melinidis*, and *B. zeicola* (= *Helminthosporium carbonum*) have been reported. Appendix 2 lists some palm taxa which are hosts of three species of *Bipolaris*. Some hosts of *Exserohilum rostratum* (= *Helminthosporium halodes*) and *Phaeotrichoconis crotalariae* are given in Appendix 3.

Prevention and Treatment: As with many diseases caused by dematiaceous hyphomycetes, fungal distribution depends on movement of conidia by wind or splashing water. Healthy plants are easily contaminated by the introduction of diseased plants into the same greenhouse or on the same bench. Splashing water, such as that produced by overhead irrigation, is ideal for dislodging conidia in large numbers and then dispersing them onto healthy tissue.

Proper environmental considerations will greatly enhance the production of vigorous, disease-free plants. Since fungal growth, sporulation, germination, infection, and subsequent disease development are highly dependent on moisture and free water, any reduction in excess moisture will curb disease levels.

Greenhouse sanitation is very important in the reduction of inoculum levels to minimize further disease development. *Bipolaris* spp. and *E. rostratum* tend to be non-specialized among palms and non-Gramineous plants. Thus badly diseased palms could serve as an inoculum source for other crops in the vicinity, or conversely, other plants may serve as the inoculum source for palm crops. Applications of chlorothalonil and carbamate fungicides such as mancozeb can provide excellent control of these leaf spots, but efficacy is drastically reduced by heavy inoculum levels or stressful conditions. Producing areca palms in full sun creates conditions in which fungicide control of these diseases is usually not successful. In addition, some young palms such as arecas are very sensitive to copper-based products and develop symptoms very similar to those created by the pathogen.

CALONECTRIA LEAF SPOT
(Cylindrocladium Leaf Spot)

Symptomology: Leaf spots caused by *Calonectria* spp. (*Cylindrocladium* asexual state) are characteristically grayish-brown, dark brown or nearly black, circular to irregular and sometimes with a dark border. Young spots are brown, circular (0.5 mm diam) and frequently surrounded by a chlorotic band approximately 1 mm wide. Expanding spots become dark brown to black with grayish-brown centers, and nearly circular to elliptical (up to 3 x 5 mm). Advanced stages of the disease are characterized by coalescing leaf spots, chlorosis and necrosis of leaflet margins and tips. Entire leaflets or leaves may be lost.

Closeup of lesions on *Howea forsterana* caused by *Calonectria theae*.

Causal Organisms: *Calonectria theae*, *C. colhounii*, and *C. crotalariae* cause leaf spots on *Howea forsterana* as described. Leaf spots caused by these three fungi are virtually indistinguishable. The distinctive conidiophores of these fungi are comprised of spore-forming phialides borne on penicillate branches which arise laterally from the main axis; the conidiophore axis is frequently very long and terminates in a swollen vesicle. The vesicles are narrowly clavate for *C. theae* and *C. colhounii*, while *C. crotalariae* is distinguished from the other two species by the production of sphaeropedunculate vesicles. Conidia of all three species are predominately 3-septate, although those of *C. theae* are narrowest. In addition, most isolates of *C. theae* produce macroconidia which are 3-14 septate, and are acutely curved to nearly right angled. Macroconidia are readily formed in water agar and are borne sympodially on simple conidiophores. Perithecia of *C. theae* and *C. crotalariae* are reddish-orange, turn brown with age, and produce asci with 8 ascospores each. Perithecia of *C. colhounii* are yellow, turn yellowish-brown with age and have asci each with 4 ascospores.

Lesions on *Howea forsterana* one month after inoculation with *Calonectria crotalariae*.

Occurrence and Species Affected: There are only a few published reports of *Cylindrocladium* spp. on palms. *Howea forsterana*, *Ptychosperma elegans* and *Washingtonia robusta* have been reported to be naturally infected by *Cylindrocladium* in Australia. Pathogenicity of *Cylindrocladium floridanum* to *H. forsterana* and *W. robusta* has been confirmed. Disease also developed on *Caryota*, *Chamaedorea*, and *Chrysalidocarpus* but only upon inoculation of wounded plants with *C. floridanum*. *Cylindrocladium pteridis* (= *C. macrosporum*) was shown to cause leaf spots on *W. robusta* in Tennessee. In Georgia, isolates of *C. pteridis* from *W. robusta*, *Eucalyptus cinerea*, and *Rumohra adiantiformis* were equally pathogenic to *C. elegans*, *H. belmoreana*, and *H. forsterana*.

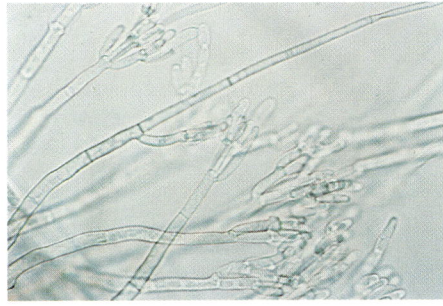

Closeup of stipe and lateral branching of conidiophores of *Calonectria theae*.

In Hawaii, *Calonectria theae*, *C. colhounii*, and *C. crotalariae* have been confirmed as pathogens of *H. forsterana* without wounding. *Chrysalidocarpus lutescens*, *Chamaedorea elegans*, and *Laccospadix australasica* were less susceptible than *H. forsterana* to *C. theae* and *C. colhounii*. Seedling blight of *Laccospadix* characterized by severe root and petiole rots has also been associated with *C. crotalariae* in Hawaii.

Diagnostic Techniques: Leaf spots and blights caused by *Calonectria* spp. closely resemble those caused by *Bipolaris* spp.

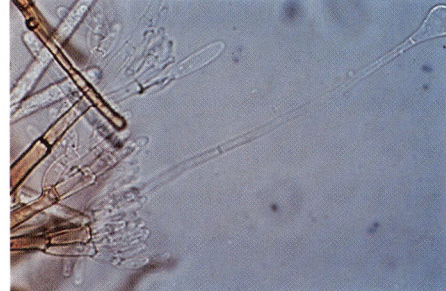

Stipe and vesicle of *Calonectria theae*.

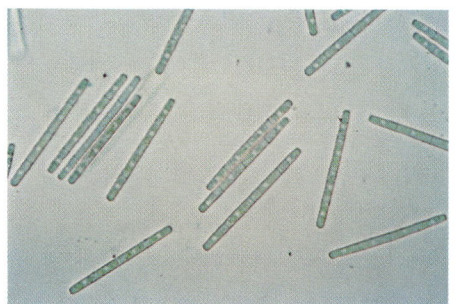

Conidia of *Calonectria theae*.

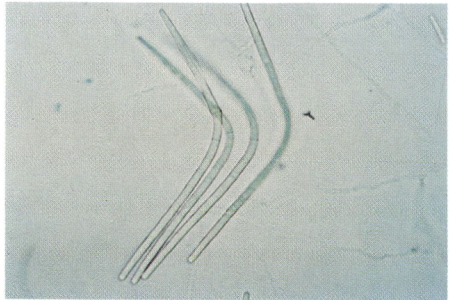

Macroconidia of *Calonectria theae*.

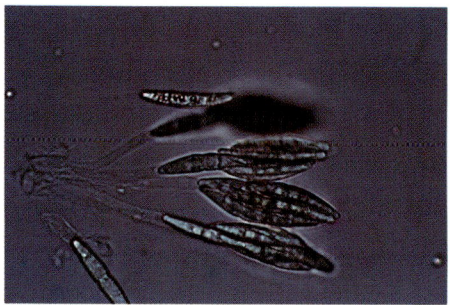

Asci with eight ascospores from perithecia of *Calonectria crotalariae*.

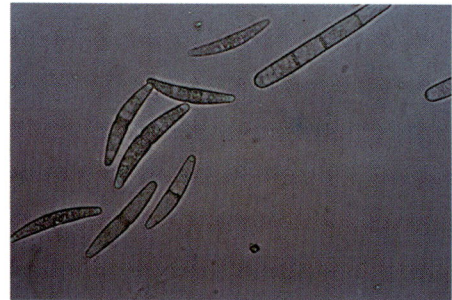

A single conidium and several ascospores of *Calonectria crotalariae*.

although *Bipolaris* lesions tend to be slightly more yellow on some palms. Microscopic examination of large spots and blights will often reveal conidia of either pathogen. In the absence of spores, incubation of diseased specimens in a moisture chamber should result in sporulation enabling ready identification of *Cylindrocladium* or *Bipolaris*.

Calonectria (*Cylindrocladium*) species can be readily isolated from leaf spots and blights, petiole spots and roots of young plants. Pure cultures of these fungi are obtained by washing leaves thoroughly with a mild soap, rinsing by rubbing under running tap water, cutting out the interface of healthy and diseased tissue from large spots or blights (or entire small spots), dipping a few second in 0.1% sodium hypochlorite, draining excess solution by touching a clean paper towel, and plating cut pieces on water agar. After incubation at about 25°C under continuous fluorescent illumination, mycelia and conidia will develop in a few days. These can be aseptically transferred to a nutrient medium of choice, although most isolates grow and sporulate well on vegetable juice (Campbell's V-8 juice) agar, or potato dextrose agar, when grown under fluorescent light.

Prevention and Treatment: Pathogenicity tests of *C. theae* and *C. colhounii* on *Howea* demonstrated that continuous high humidity was required for disease development. Since there is a high moisture requirement for disease development, it follows that moisture control is important to disease prevention. Any cultural practice that reduces free moisture on leaves (e.g., increased spacing among plants, covered greenhouses, drip irrigation, etc.) will reduce disease incidence and spread.

Conidia are also produced under moist conditions on leaf surfaces. These spores spread to healthy plants by splashing water, and by the use of contaminated potting media, tools, etc. Potential spread is greatly increased with the production of airborne ascospores, which are forcibly discharged from perithecia and carried by air currents. Unlike conidial spread, dissemination of ascospores is not dependent on free moisture although ascospore discharge is induced by a rapid increase in relative humidity. Prompt removal and destruction of infected leaves, especially blighted leaves, before perithecia are formed is crucial to disease control. Reduction of inoculum sources greatly increases cost-effectiveness of chemical treatments. Benomyl and maneb are effective against these *Calonectria* species.

CATACAUMA LEAF SPOT
(Tar Spot)

Symptomology: Infection can occur on pinnae, rachides, and peduncles. Lesions are diamond-shaped with the long axis parallel with the venation of the leaflet. Lesions have a crusty, wart-like texture and measure 4-6 mm x 2-4 mm. A yellow halo surrounds the infected spot. Disease incidence can be quite high, such that large areas of tissue become necrotic.

Causal Organism: *Catacauma sabal* is the cause of tar spot on palms. Although many species of the genus *Catacauma* have been re-examined and placed in the genus *Phyllachora*, the disposition of *Catacauma sabal* is uncertain. The fungus grows and ramifies below the epidermis on either leaf surface but does not extend into the mesophyll. The dense layer of mycelia (stromata) is dark brown to black in color and measures 2-3 x 1.5 mm. Within the stromata, the sexual stage forms in several large locules measuring 240-250 x 93-120 µm. The spores are born in groups of eight in a membrane-bound structure that is 65-85 x 35-43 µm in size. The individual spores are colorless, 1-celled, and measure 25-26 x 7-12 µm. The asexual stage, an acervulus, is seen rarely and produces conidia that are colorless and measure 4.5-6 x 2-3 µm.

Two other species (*Catacauma mucosum* and *C. torrendiella*) are reported from South America. The status of these species as pathogens is unclear.

Occurrence: This pathogen is reported from the Dominican Republic, Mexico, United States (Florida, Georgia, Texas), and the West Indies.

Species Affected: *Acoelorrhaphe wrightii, Livistona chinensis, Sabel causiarum, S. etonia, S. mexicana, S. minor, S. palmetto, S. umbraculifera, Syagrus romanzoffiana,* and *Washingtonia robusta* have been reported as hosts of *Catacauma sabal*. *Catacauma mucosum* is reported from *Butia, Cocos,* and *Syagrus* spp., while *C. torrendiella* is reported from *Cocos nucifera*.

Diagnostic Techniques: Since this pathogen is difficult to culture, direct microscopic examination is the preferred diagnostic technique. Lesions should be sectioned along both axes in an effort to section through the locules. Since the timing of spore maturity is not known, most sectioning attempts will reveal empty, carbonaceous-looking locules. The occasional section will reveal the sexual spores and confirm the diagnosis.

Prevention and Treatment: With so little of the infection cycle of this pathogen documented, the use of either sanitation or fungicides is haphazard at best. The removal of severely infected fronds will reduce the inoculum left on the plant, but the critical timing for this activity is not known. Incidence of this disease seems to be higher for those palms grown under less than optimal light levels. Planting sites with poor air circulation and overhead irrigation will likely favor disease severity by stimulating spore release, dissemination, and infection. Use of such fungicides as Bordeaux mixture or copper-based products will aid in the control of this problem. Fungicide efficacy trials conducted in Brazil against a related fungus on palm (*C. torrendiella*), demonstrated effective leaf spot control with benomyl, carbendazim, carboxin, chlorothalonil, cuprous oxide, maneb, oxycarboxin, thiabendazole, thiophanate methyl, triadimefon, or tridemorph. Best timing of fungicide use is unknown.

Catacauma leaf spot of *Chamaerops humilis*.

Catacauma leaf spot of *Acoelorrhaphe wrigthii*.

Closeup of Catacauma leaf spot of *Chamaerops humilis*.

Locules and extruded asci of *Catacauma sabal*. (Courtesy M. Gooch)

DAMPING-OFF

Damping-off of *Roystonea* seedlings.

Damping-off of *Livistona* seedlings.

Diamond scale on *Washingtonia filifera*.

Ascostroma of *Sphaerodothus neowashingtoniae*.

Cross-section of ascostroma of *S. neowashingtoniae*. (Courtesy M. Murphy)

Symptomology: The most common early symptoms of damping-off are either poor germination which is referred to as pre-emergence damping-off or loss of seedlings after germination (post-emergence damping-off). Seeds or seedlings are mushy and collapse. An overall poor stand results which makes repotting or discarding necessary.

Causal Organism: A variety of organisms cause damping-off but the most common culprits are fungi such as *Fusarium, Phytophthora, Pythium,* and *Rhizoctonia* spp.

Occurrence and Species Affected: These organisms usually have wide host ranges and are frequently found affecting many plant species in the same nursery. It is likely that all palms produced from seeds are subject to damping-off.

Diagnostic Techniques: Isolation of the causal organism(s) is usually easy if several semi-selective culture media are used. Sections of rotted roots, stems and seeds can be surface-disinfested using standard methods and plated on culture media such as potato-dextrose agar amended with an antibiotic (to reduce bacterial contaminants) and one for pythiaceous fungi. Colonies are produced at different rates, with *Rhizoctonia* and *Pythium* usually growing faster than *Phytophthora* and Fusarium. Keep in mind that a damping-off problem may be caused by one of more of these fungi as well as a variety of others such as *Cylindrocladium* or *Sclerotium* spp.

Prevention and Treatment: Always use clean palm seed, new potting media, clean pots, and whenever possible, grow plants on raised benches to limit exposure to native pathogens in the soil under the pots. Preventive drenches with a variety of fungicides are often chosen by growers as extra insurance against damping-off of expensive seeds.

DIAMOND SCALE

Symptomology: This disease is characterized by the appearance of the black, diamond-shaped fruiting structures on the leaf blades and petioles. The leaves turn chlorotic, senesce prematurely and die.

Causal Organism: *Sphaerodothus neowashingtoniae* is the causal organism of diamond scale. Numbers of ascostroma on a rachis may range from several to thousands. The ascostroma appear shallow on the petiole while on the leaf, stroma form on both leaf surfaces of the blades. Ascospores are released by weathering of the stroma and are produced in such numbers that the leaf may appear black. Germination of spores around the stroma rings it with new lesions and eventually many small fruiting structures.

Sphaerodothus neowashingtoniae produces chesnut brown to black ascospores measuring 56-68 x 30-36 μ. In shape they are oblong-elliptic, smooth and somewhat flattened on one side with a narrow oblong, longitudinal depression in the middle, suggesting in appearance a date seed. Free, mature spores frequently collapse, becoming saddle-shaped.

Occurrence and Species Affected: Diamond scale of *Washingtonia filifera* has been reported from the United States (California and Arizona).

Diagnostic Techniques: The occurrence of the black, diamond-shaped ascostroma is diagnostic. These structures range in size from 1 to 2 mm to several centimeters in length with the long axis parallel to the leaf veins.

Prevention and Treatment: Removal of infected foliage and use of a protective fungicide may help although appropriate timing of the treatment is unknown.

Ascospores of *S. neowashingtoniae*. (Courtesy M. Murphy)

FUSARIUM WILT

Symptomology: Affected trees are characterized by the death of fronds in patterns that differ from normal. Fronds on one side of the tree may die more rapidly than the others producing a lopsided appearance, fronds may die more rapidly than usual from the lower fronds upward or a ring of fronds may die with green fronds below and above them. An individual frond may have spines and pinnae die progressively from the base to the tip while the pinnae on one side are still healthy on the opposite side of the frond. Individual pinnae may show watersoaking along the vascular elements. A dark-brown streak often appears on the bottom of the rachis. Upon splitting of the rachis, the vascular elements will often be streaked dark brown as in other Fusarium wilts and fungal hyphae can be observed in cross-sections of the vascular tissues. The disease is often found in association with pink rot caused by *Gliocladium vermoeseni* which attacks the tree when it is stressed by the primary pathogen *Fusarium oxysporum*.

Causal Organism: *Fusarium oxysporum* is the cause of Fusarium wilt on palms. The pathogen produces an appressed orange to salmon-pink colony on potato-dextrose agar. Conidiophores produce both macro- and microconidia. Macroconidia have 2 or 3 septations and measure 30-50 x 3-5 μm. Microconidia are ovate to elliptical and measure 3-17 x 2-8 μm. Chlamydospores are 2-11 μm in diameter and may be terminal, intercalary, or form from macroconidia. Optimum temperature range for growth is 24-30°C.

Fusarium wilt of *Phoenix canariensis*.

Occurrence: This disease has been reported from California only. Diseases with similar symptoms have been reported from France, Italy, Japan and Australia. Fusarium wilt of *P. canariensis* is similar to Bayoud disease of date palm (*P. dactylifera*) in Morocco and Algeria which is caused by *F. oxysporum* f. sp. *albedinis*.

Species Affected: *Phoenix canariensis*, *P. reclinata*, and seedlings of *P. dactylifera* are currently known to be susceptible to *F. oxysporum*.

Diagnostic Techniques: The pathogen can be cultured from affected tissue using potato dextrose agar containing 1 to 5 ppm benomyl to suppress the growth of *Gliocladium*. The occurrence of the above described symptoms and culture of the organism is diagnostic.

Fusarium wilt of *Phoenix canariensis* after inoculation with *F. oxysporum*.

Prevention and Treatment: The pathogen is soil- and

One-sided death of pinnae on *Phoenix canariensis* due to Fusarium wilt.

Longitudinal-section of *Phoenix canariensis* petiole with brown vascular streaking (*F. oxysporum*) and black rot (*G. vermoseni*).

Conidia of *Fusarium oxysporum*.

Ganoderma butt rot on *Phoenix reclinata*.

Ganoderma butt rot on *P. reclinata*.

water-borne. In California it has been shown to be transmitted during pruning operatiosns by chain saws and hand saws. There is no treatment; all infected trees eventually die. Transmission of the pathogen can be prevented by using pruning saws that can be and are thoroughly sterilized between pruning trees. Sterilization can be accomplished by a 5 minute dip in a 2.5% solution of sodium hypochlorite. Do not use chain saws on trees affected by Fusarium wilt due to the difficulty of sterilization.

GANODERMA BUTT ROT
(Basal Stem Rot)

Symptomology: The initial symptom of butt rot of palms is withering and drooping of older fronds. As these fronds collapse, the pinnae may roll back along the rachis and the entire frond droops parallel to the trunk. Fronds do not snap off at the rachis. New growth slows, decreases in size, and develops a pale-green to yellowish cast. Young expanded fronds may exhibit no color shift for several years. As growth slows, several unopened spears may form in the crown. Flowering also will slow and ultimately stop on infected palms. As the older canopy continues to die, younger leaves may exhibit nutrient deficiency symptoms, progressive periods of wilt, and tip necrosis. Finally, only one or more spears remain in the bud. These discolor and die. Depending on the location of the infection in the tree, the head of the palm may fall off or the trunk collapse. Palm death may take 3-4 yrs depending on age and environmental conditions.

Other symptoms are apparent if the lower trunk is dissected into the root zone. Although the outer trunk tissues are solid, affected palms have a hollow sound when tapped. Within the trunk is a zone of dark brown infected tissue that is surrounded by a narrow, darkly stained band. Surrounding this discolored zone is the advancing margin of the pathogen. The tissue in this area is yellowish, rotted, with extensive mycelia present. Depending on the point of invasion, the roots may be extensively decayed. Invaded roots contain fungal hyphae throughout the brown, decayed cortex. The conductive tissues (stele) appear dark brown to black in color and are no longer functional. On the lower trunk adjacent to this area, a bracket (basidiocarp) is often present soon after early decline symptoms begin.

Causal Organism: *Ganoderma zonatum* (Synonyms = *Ganoderma sulcatum*, *Polyporus lucidus* var. *zonatus*, *Ganoderma tumidum*, and *Ganoderma applanatum*) is the cause of Ganoderma butt rot of palms. The basidiocarp of *G. zonatum* is woody, kidney-shaped, and brightly colored. Basidiocarps emerge annually from root or trunk tissue, usually after initial symptoms of palm decline are evident. Young basidiocarps are swollen white protuberances and as these mature, the upper surface will develop a brown color (variable) with a somewhat shiny finish. The basidiocarp may expand to a diameter of 30-40 cm at the widest point with a maximum thickness of 9 cm at the base or point of attachment to the palm. The upper surface may be knobby to rippled in texture. The basidiocarp exhibits a swollen edge, often revealing the pure white, lower spore-bearing surface that, at maturity, appears minutely pored. Although previous-year basidiocarps will persist on the palms they are not fertile.

Cutting through a basidiocarp will reveal three layers of tissue. The upper cuticle-like layer is only 20 μm thick with a yellow color. The inner layer may extend through one third to one half of the depth of the bracket and is a bay color. The lower layer is the fertile tube layer that may span one half to two thirds the bracket thickness and is walnut brown in color. Each tube ends in a minute pore as observed on the lower layer of the basidiocarp. From these pores emerge the sexual spores that are light yellow to tan in color, ellipsoid in shape, and measure 9-16 x 5.5-9 μm.

Occurrence: Reports of *G. zonatum* are somewhat vague due to its earlier identification as *G. sulcatum, G. lucidum,* or *G. applanatum*. This pathogen is reported from the southeast United States (Alabama, Florida, Georgia, and South Carolina), throughout tropical Africa, and Argentina. Distribution is believed to extend throughout much of South and Central America and into Asia. Related species that infect palms include *G. boninense* reported from Australia, Japan, Indonesia, Malaysia, Philippines, Samoa, Sri Lanka, and Tasmania. By contrast, *G. tornatum* is widely reported throughout the tropics. It occurs from the southern tip of Africa, up the western shore of Canada in the Pacific, as well as from India and Pakistan. It is not reported from southern Europe or north of the Sahara in Africa and distribution in South America is unknown.

Species Affected: *Ganoderma zonatum* is reported only from the palm family with the exception of one report of this species recovered from a *Eucalyptus* sp. See Appendix 4 for a partial list of palm hosts of *G. zonatum*. Hosts of other *Ganoderma* spp. are less defined. *Ganoderma boninense* has been reported on *Areca* spp., *Cocos nucifera, Elaeis guineensis,* and *Livistona chinensis* var. *subglobosa*. *Ganoderma tornatum* is reported from the palm family in the tropics and many other host species.

Diagnostic Techniques: This disease can be diagnosed by the presence of the sessile, kidney-shaped basidiocarp that is characterized by a white lower surface and a shiny, brown upper surface. If the basidiocarp is mature, the basidiospores are ellipsoid, golden yellow to light brown and measure 9-16 x 5.5-9 μm. No rhizomorphs are known to occur with this pathogen.

Microscopic examination of hyphae from a mycelial felt underlying the root epidermis or the external trunk surface will reveal typical hyaline hyphae (4-7 μm in diameter) with clamp connections. This pathogen will grow from spores or internal basidiocarp tissues at 25-30°C on malt extract agar. After 3-4 weeks in culture, the white mycelial mat will develop a yellowish-tan color. The culture exhibits a ribbed margin, a fruity odor, and aerial mycelial growth. Mycelia do not produce chlamydospores but do produce numerous aerial, staghorn-like hyphae with numerous side branches.

Although antisera have been developed for the related species, *G. lucidum,* this diagnostic technique is not available for *G. zonatum*.

Prevention and Treatment: Various preventive and curative measures for butt rot control have been tried but with limited success. In landscape situations, avoid the initial

Ganoderma butt rot on *Syagrus romanzoffiana*.

Upper and lower surface of mature sporophore of *Ganoderma zonatum*.

Cross-section of *Phoenix dactylifera* killed by *Ganoderma zonatum*.

Emerging sporophore of *Ganoderma zonatum* on a *Phoenix canariensis*.

Mature sporophore of *Ganoderma zonatum* on a palm stump.

Sporophores of *Ganoderma zonatum* from two seasons.

Mature sporophore of *Ganoderma zonatum* on palm stump hidden in grass.

Fruiting bodies of *Graphiola phoenicis* on *Sabal palmetto*.

establishment of this pathogen. During landscape design, place palms in sites where trunks will not be prone to injury. Space palms adequately to prevent spread of Ganoderma within a site through below ground contact between root systems of infected and healthy palms. Exercise care during maintenance activities to avoid wounding of the palm trunk with mowing and/or trimming equipment. The loss of a palm to lightning, transplant shock, or disease should trigger quick specimen removal and destruction by burning or burial off site since *Ganoderma* is an excellent colonizer of dead tissue.

Sanitation is the major emphasis in *Ganoderma* treatment programs world wide. Promptly remove root systems, stumps, and trunks of dead palms in the landscape or nursery. This will prevent future airborne spread of this pathogen from basidiocarps formed on these trees. Root system removal will suppress soil movement of this pathogen toward adjacent susceptible species. If a fallow period cannot be observed, the planting site should be sieved free of root fragments, filled with clean soil, and fumigated (if possible) prior to palm replacement. Products containing metam sodium or methyl bromide/chloropicrin have utility in site fumigations where labeling allows legal use.

Fungicides have been examined for their possible use in controlling butt rot. Although products containing either mercury, thiram, or tridemorph have been effective in inhibiting the fungus in culture, field use of these products has failed. Limitations to field effectiveness include 1) the lack of a presymptomatic, early detection method for Ganoderma infection, 2) the non-systemic nature of some fungicides toxic to Ganoderma, and 3) incomplete distribution of systemic fungicides within infected root and trunk tissues of palms. Fungicides and other chemicals have been examined for their use in stump treatment to suppress Ganoderma reproduction. Stump treatments with creosote, copper-based fungicides, and tridemorph have been ineffective. Currently, hand removal of reproductive brackets of Ganoderma is recommended until the stump can be removed.

GRAPHIOLA LEAF SPOT
(False Smut)

Symptomology: Symptoms of false smut are obscure in contrast to the sign of the causal fungus. Small yellow to brown spots develop on pinnae (both surfaces), rachides, and leaf bases. These spots swell and eventually rupture as the reproductive sori of the fungus emerges from below the leaf epidermis. Incidence of leaf lesions is often greater on the upper surface of pinnae and on those pinnae closest to the frond base. Because of the 10-11 month incubation cycle for this pathogen, disease incidence generally occurs on 2-yr and older fronds. High densities of lesions can result in premature senescence of older fronds. The normal 6-8 yr life of date palm fronds has been reduced by false smut to 3 years in commercial date growing areas of the world.

Causal Organism: Since the first description of *Graphiola phoenicis* from palm in 1823 by Mougeot, it has been classified as both a slime mold and a true fungus. It is now considered to belong in its own order (Graphiolales) within the class Heterobasidiomycetes being related to the smut fungi.

The most salient feature of this pathogen is the wart-like, cup-shaped, black reproductive structures (sori) that rupture through both leaf surfaces. Sori emerge from the interveinal tissue or along veins in the pinnae and may be solitary or clustered and reach a density of 20 or more per cm^2 of leaf. Sori range in size from 0.5-2 mm in diameter. From these a column (1-1.5 mm thick) of spore-bearing hyphae will emerge and extend 1-2 mm beyond the sorus. This column of hyphae is believed to be the site of nuclear fusion and reduction division prior to the formation of two spore stages. At sexual maturity, the column of fertile hyphae will appear yellowish as numerous spores are released to begin a new disease cycle. These spores are 1-celled, oval, and measure 3-6 μm in diameter. They can germinate by either the emergence of septate hyphae or budding. The mycelia infect through the stomata of the leaf growing inter- and intracellularly over the next 10 months until sorus emergence occurs.

Two other species, *G. congesta* and *G. thaxteri* are reported from the southeastern United States. The status of these species is not clear at this time.

Occurrence: This fungus is found worldwide in tropical to subtropical areas where rainfall is plentiful. *Graphiola phoenicis* is not very damaging in arid areas. Reports of this disease originate from Algeria, Argentina, Austria, Belgium, Brazil, Canary Islands, Columbia, Cuba, Dominican Republic, Egypt, Fiji, France, French Guiana, Germany, Great Britain, Greece, India, Italy, Jamaica, Japan, Mali, Mauritania, Netherlands, Niger, Peru, Senegal, Tabago, Trinidad, United States (Arizona, California, Florida, and Puerto Rico), Uruguay, Venezuela, West Indies, and a general report from the Scandinavian region.

Graphiola congesta and *G. thaxteri* have been reported only from the southeastern area of the United States.

Species Affected: The genus *Graphiola* has a host range limited to the Arecaceae. See Appendix 4 for a partial list of hosts of *Graphiola phoenicis*. *Graphiola congesta* is reported only from *Sabal palmetto* while *G. thaxteri* is reported from *Sabal megacarpa*.

Diagnostic Techniques: The presence of the sori on foliage is the diagnostic sign of this fungal pathogen whether spore-bearing hyphae are present within or not. Immature sori can be stimulated into reproduction by placement in a high humidity chamber for several days. The yellowish fascicle of hyphae is again diagnostic for this pathogen. The spores can be cultured at 25°C on a variety of growth media including potato sucrose agar, oatmeal agar, and malt extract agar. Growth is slow (8-10 mm in 10 days) with the fungus appearing yeast-like, flat, with a pale orange to pink color and a smooth margin. Individual cells are cylindrical, measuring 5-7 μm x 1.5-2 μm. Relative humidity must remain ±95% to maintain colony growth.

Prevention and Treatment: Sanitation of severely diseased fronds is usually an adequate control for ornamental palms although frond removal should be done in moderation since research with date palms indicates that frond removal can result in a serious loss in

Foliar infection of two pinnae of *Phoenix dactylifera* with *Graphiola phoenicis*.

Emergent spore-bearing hyphae of *Graphiola phoenicis*.

Closeup of emergent spore-bearing hyphae of *Graphiola phoenicis*.

Lethal yellowing on Jamaican Tall *Cocos nucifera*.

Flag leaf death on Jamaican Tall *Cocos nucifera* with LY.

Sudden wilt symptoms on Malayan Dwarf *Cocos nucifera* with LY.

Foliar yellowing of *Cocos* with LY. (Courtesy N. Harrison)

vigor. Fungicides have been evaluated for false smut control on date palms in India. Products containing either mancozeb, cupric hydroxide, cupric hydroxide + maneb, or copper oxychloride + maneb + zineb controlled this disease if 3-4 sprays were applied on a 15-day schedule after sporulation was first observed. Fungicide product labels should be examined for their legal use on ornamental palms. Resistance to false smut is available for commercially grown date palm cultivars including Adbad, Barhee, Gizaz, Iteema, Kustawy, and Rahman which are tolerant or resistant to this pathogen. Resistance in other palm genera is not unknown.

LETHAL YELLOWING DISEASE
(Awka Disease, Cape St. Paul Wilt, Kaincope Disease, Kribi Disease, and Pudricion Del Cogollo)

Symptomology: Lethal yellowing (LY) symptoms are variable among palm genera. The progression of symptoms for coconut palms is well defined whereas LY symptoms on other palms can be grouped by the type and order of foliar discoloration.

1. Coconut palm - The initial symptom is premature nut fall or "shelling" on mature palms. Nuts fall first from the mid-crown (water nuts), followed by younger nuts, and lastly mature nuts. Fallen nuts exhibit a dark-brown to black, water-soaked zone at the stem end. Secondary symptoms involve inflorescence discoloration. Normal flowering stalks should appear creamy white to yellow as they emerge from the spathe. Palms infected with LY will exhibit flowering stalks with black or dark-brown male flowers on the tips. In advanced stages of LY, the spathe may discolor and fail to open. Flower necrosis is a definitive symptom for LY. Foliar discoloration represents the third stage of symptom development. In general, one or more of the oldest, lower fronds turns yellow, then brown, and hangs parallel to the trunk. Immature fronds yellow (remaining upright) and then brown after bud death. Death of the bud occurs when one-half to two-thirds of the canopy has yellowed. Bud death is quickly followed by decay caused by insect and secondary microbe invasion. In Malayan Dwarf coconuts, early symptoms include browning rather than yellowing of the foliage and an overall wilt of the palm. Leaflets of infected Malayan Dwarf coconuts are folded about the midvein and the entire frond appears noticeably limp. In other coconut cultivars leaflets and fronds retain their turgidity.

2. Palms with coconut LY-like symptoms - Such palms as *Caryota mitis*, *Corypha elata*, *Dictyosperma album*, *Hyophorbe verschaffeltii*, *Pritchardia* spp., *Syagrus schizophylla*, and *Trachycarpus fortunei* exhibit early LY symptoms like those of coconut. Development of yellow flag leaves is more common in these genera, while the period of leaf yellowing is variable. Dead lower leaves may be held onto the trunk like a collapsed umbrella. In *Pritchardia* spp., spear leaf death is the first observable symptom, occurring before immature nut drop.

3. Non-yellowing palms - *Borassus flabellifer*, *Chrysalidocarpus cabadae*, *Phoenix* spp., and *Veitchia* spp. exhibit frond browning rather than bright yellowing. Although nut drop and inflorescence discoloration develop similarly to coconut palms, frond discoloration

is not as distinctive. The first evidence of foliar symptoms is a browning along the margin of pinnae. Unopened inflorescences may be twisted. As with *Pritchardia* spp., the *Phoenix* spp. develop spear leaf death as the first symptom. *Borassus* discoloration is rusty yellow, while *Phoenix* spp. discolor to a dusty gray color and *Veitchia* fronds turn rusty brown in color.

Causal Organism: In the early 1970's, mycoplasma-like organisms (MLOs) were found associated with 'Jamaican Tall' coconut palms symptomatic of LY. Shortly thereafter, similar microbes were observed in the conductive tissues of 21 palm species other than coconut. The MLOs of palm (to date) have not been cultured and thus have not been definitively named. The MLOs are variably shaped, minute wall-less microbes, that possess a tri-laminar unit membrane, ribosomes, DNA, but no organized nucleus. Their size and shape are not confirmed, but electron microscopy would indicate both spherical and filamentous components that possess diameters ranging from 1.2 μm to 500 μm. MLOs have been found in the phloem tissues of LY-plants but not healthy material. Palm insect surveys in LY-affected and unaffected areas have demonstrated a 40x greater incidence of the planthopper, *Myndus crudus* Van Duzee, in the LY-affected zone. This planthopper was also found to be the most abundant leaf-feeding insect associated with coconut palms in both Florida and Jamaica where LY was present. In addition, *M. crudus* was recovered from all but the most rare, LY-susceptible palm species in Florida.

Occurrence: Distribution is somewhat vague due to the variable symptomology among palm species and the present inability to definitely diagnose this disease. This disease has been recorded under a variety of names depending on the location. Literature indicates that LY first occurred in the Cayman Islands in 1834. Since then, LY has occurred in the Bahama Islands (1946-"unknown disease"), Cuba (1870's-Pudricion del cogollo), Dominican Republic (1925), Haiti (1920's-"unknown disease"), Jamaica (1870's-bronze leaf wilt), Mexico (1982-northeast Yucatan peninsula) and the United States (1978-Rio Grande Valley in Texas, 1971-the lower east and 1983-the west coast of Florida, and the Florida Keys-1955). In western Africa this disease is reported from Cameroon (1937-Kribi disease), Ghana (1932-Cape St. Paul wilt), Nigeria (1951-Awka disease), Togo (1932-Kaincope disease), and is suspected in Tanzania (1978-based upon symptoms). The Kerula and Karnataka states in India have also reported LY in 1977.

Species Affected: Appendix 5 gives a list of some palms reported as hosts of the lethal yellowing pathogen and others which are felt to be resistant to one degree or another. The status of non-susceptible palm species to LY is subject to change. Observations of a few specimens for some of the palm taxa may represent disease escapes. Other taxa are believed immune, but their native range does not correspond to those centers of LY activity and thus this germ plasm has not been adequately evaluated.

Diagnostic Techniques: Diagnosis of LY is based chiefly on symptomatology in the field. The most precise diagnosis of LY is dependent on transmission electron microscopy. This technique is both expensive, time and labor intensive, and not widely available to

LY symptoms on *Phoenix dactylifera*. (Courtesy N. Harrison)

LY symptoms on *Veitchia merrilli* (front) and Jamaican Tall *Cocos nucifera* (rear). (Courtesy N. Harrison)

LY symptoms on *Caryota rumphiana*. (Courtesy N. Harrison)

Necrosis of immature inflorescences of *Phoenix dactylifera* due to LY. (Courtesy N. Harrison)

Cross-sections of necrotic inflorescenses. (Courtesy N. Harrison)

Nut drop on Panama Tall *Cocos nucifera* due to LY. (Courtesy N. Harrison)

diagnostic facilities. Preferred tissue for evaluation by microscopy includes the phloem tissues of young leaf bases, spear leaves, and inflorescences. Generally, MLOs can only be observed in 2-13% of the functional sieve elements in this young tissue. Exceptions to this MLO frequency occur in such palms as *Phoenix canariensis* and *Trachycarpus fortunei* where over 50% of the vascular bundles will contain MLOs. In spite of the sensitivity of electron microscopy, negative results may be falsely interpreted. The LY MLOs exist at such low levels in plant tissues and are so random in distribution that they may be missed during microscopy--especially with so small a sample of tissue examined.

Prevention and Treatment: Specific geographical areas have quarantine procedures in place to deal with the importation of known susceptible palm genera and/or species. As palms are purchased and planted, common sense should dictate that a diversity of palm germ plasm be utilized to avoid the catastrophic results experienced with LY in Florida in the 1970's through the mid 1980's where high percentages of all coconut palm plantings consisted of the susceptible 'Jamaican Tall' variety.

With considerable evidence supporting the role of *M. crudus* as a vector of LY in Florida, control of LY by vector management may be useful. Research has shown that such pesticides as dimethoate and diazinon significantly reduce the rate of spread of LY compared to unsprayed control plots. However, the degree of LY control is not high enough to support repeated pesticide applications in either the ornamentals landscape or palm plantations at this time.

Curative control efforts have centered around the application of various antibiotics, fungicides, metal salts, and insecticides. Therapeutic benefits were demonstrated in LY-symptomatic palms after the use of the antibiotics oxytetracycline HCl or tetracycline HCl. Field rates of these antibiotics are 1-3 g(ai)/tree on a 4-month treatment schedule. Palms are generally unaffected by the applications of these antibiotics at rates as high as 20 g/tree.

Antibiotic treatment should begin as early in symptom expression as possible or be used preventively when LY is known to occur in the area. Symptomatic palms with >25% of the canopy yellowed should be removed as they are unlikely to respond to antibiotic treatment. The preferred method of application is by liquid injection (>2 inches deep) into the trunk. Four methods are described in Appendix 6. Regardless of the method selected, certain precautions should be followed. The reservoir should be made from either brass, stainless steel or plastic. Coconut palms grown for milk and/or meat consumption can be safely harvested 30 days after treatment without detectable levels of antibiotics in either meat or milk.

The use of host resistance represents the most stable, long-range control measure for LY. Such coconut varieties as the Indian Dwarf, Ceylon Dwarf, Fiji Dwarf, and King have demonstrated high resistance to LY with fewer than 10% of trees developing symptoms. Other selections like Bougainville Tall, Kar Kar Tall, Malayan Dwarf, Malayan Tall, Markham Valley Tall, and Panama Tall, and hybrid Maypan exhibit moderate levels of resistance. Green forms of Malayan Dwarf appear very susceptible with losses of 50-70% while losses of golden froms have usually been less than 30%. Aside from

Closeup of nuts affected by LY. (Courtesy N. Harrison)

Injection site for antibiotic treatment for LY affected *Cocos*. (Courtesy N. Harrison)

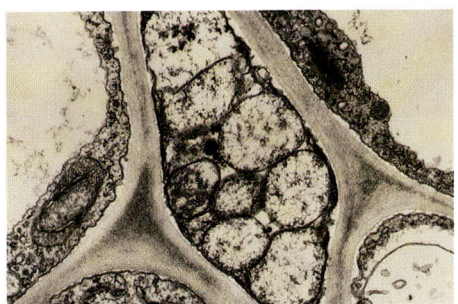

Electron micrograph of mycoplasma associated with LY. (Courtesy N. Harrison)

Myndus crudus planthopper vector of LY mycoplasma. (Courtesy J. DeFilippis)

Palm mosaic virus on *Washingtonia robusta*. (Courtesy D. Mayhew)

Palm mosaic virus on *Washingtonia robusta*. (Courtesy S. Koike)

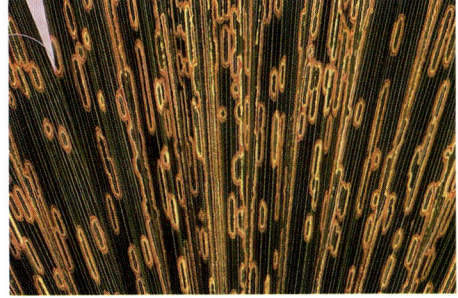

Palm mosaic virus on *Washingtonia robusta*. (Courtesy S. Koike)

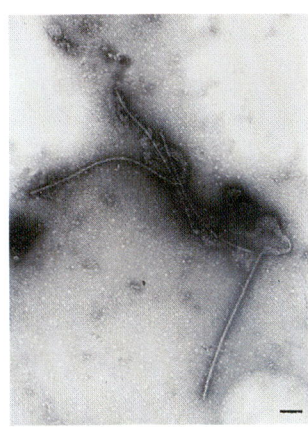

Electron micrograph of palm mosaic virus particle. (Courtesy D. Mayhew)

Pestalotiopsis leaf spot on *Syagrus romanzoffiana*.

Closeup of Pestalotiopsis leaf spot on *Phoenix canariensis*.

Pestalotiopsis in crown tissue of *Phoenix roebelini*.

Closeup of Pestalotiopsis lesions on palm rachis.

Cocos nucifera, many other palm genera are being screened for resistance or immunity to LY to provide alternative ornamental palm species for landscape plantings.

MOSAIC

Symptomology: Infected plants exhibit a bright mosaic on leaves. The mosaic symptoms disappear on older leaves while bright yellow rings and/or line patterns develop on this older canopy. New leaves are smaller than normal and plant vigor is reduced.

Causal Organism: Flexuous rod-shaped virus particles have been observed in symptomatic tissue. Particles measure 686 nm in length with a 13 nm width (mean dimensions). Transmission electron microscopy reveals pinwheel inclusion bodies throughout the leaf mesophyll. Neither mechanical transmission, vector involvement, nor virus purification has been demonstrated.

Occurrence: This disease has been reported from California. Another rod-shaped virus has been reported from the Philippines and Malaysia.

Species Affected: *Washingtonia robusta* is the only known host of this disease. The second virus report from the south Pacific involved *Cocos nucifera*, *Corypha elata*, and *Livistona rotundifolia*.

Diagnostic Techniques: The viral nature of this disease can be determined by transmission electron microscopy of leaf dips of freshly cut palm leaves in 2% phosphotungstic acid (pH = 8.0). No inclusion bodies have been observed with light microscopy.

Prevention and Treatment: Rogue and destroy infected plant material.

PESTALOTIOPSIS LEAF SPOT

Symptomology: Pestalotiopsis leaf spot starts as tiny black spots which enlarge to form approximately circular lesions up to 2 mm in diameter. When mature the lesions are almost white with a black border and black spot in the center. Lesions may form on leaf margins or in blades or on the rachis tissues.

Causal Organism: *Pestalotiopsis palmarum* has been recovered from numerous palms in Florida.

Occurrence and Species Affected: This disease has been reported on *Elaeis guineensis* throughout the world as well as from many palms grown in Florida. Appendix 4 contains a partial list of reported hosts of *P. palmarum* from Florida.

Diagnostic Techniques: This fungus is readily recovered from lesions by standard surface-disinfestation and plating methods on potato-dextrose agar. Colonies are initially white and fluffy and later covered with black conidia.

Prevention and Treatment: Preventive applications of broad-spectrum fungicides may aid in disease control. Elimination of overhead watering, decreased exposure to rainfall and promotion of leaf drying through appropriate irrigation timing and plant spacing are recommended for control.

PHYTOPHTHORA DISEASES

Diseases caused by *Phytophthora* species on different palm crops include seedling blights and damping-off; trunk, crown and root rots; leaf spots, blights, and petiole rots; nut drop; and apical tip, bud, or heart rot followed by the death of the plant.

Symptomology: *Phytophthora nicotianae* (= *P. parasitica*) causes seedling blight of *Chrysalidocarpus lutescens*. The disease begins as necrotic brown flecks, which expand to angular, irregularly shaped leaf spots, and finally into large, blighted areas with occasional yellow borders. Diseased tissue is initially grayish-black to grayish-green but turns tan to brown with blight progression. Leaf lesions may be as long as 40 mm, or up to 1/3 the leaflet length. Infections of the emerging leaf spreads into unfurled leaves eventually killing the terminal bud and the plant.

Bud decay affecting several species of palms in India is caused by *Phytophthora palmivora*. Incidence of this disease is correlated with a 2-6 month period following hurricanes or severe storms. Symptoms begin as a pale-green discoloration of the spear leaf and one or more of the newest leaves. Unfolding pinnae may exhibit dark-brown lesions from infection that has occurred in the spear. Successively emerging leaves may exhibit a 'bitten leaf' appearance due to damage incurred in their primordial state. The spear rots at the base and pulls out easily. Leaf bases exhibit a brown, necrotic lesion while inner leaf bases may be covered with a white mycelial growth. Invaded tissue appears oily and light brown to yellow in color. Below the rotted bud, there is an expanse of purple to pink discoloration in the tissue with a dark brown margin. Rapid disease development can result in crown death while the lower canopy remains green for 6-12 months. Slow disease development will produce a progressive frond yellowing, browning, and collapse from the lower canopy upward until the crown is dead. Fronds will often fall from the trunk leaving the dead trunk naked.

Trunk and collar rots of *Washingtonia robusta* are caused by *P. palmivora* as well as by an undetermined *Phytophthora* sp. resembling *P. drechsleri*. The initial symptom is a decrease in palm vigor, as evidenced by a pale-green discoloration of the spear leaf. Progressive wilt develops in these leaves, followed by desiccation and death and the bud rots and pulls out easily. The base of the bud smells foul and is darkly discolored. Tan to dark-brown lesions form on the leaf bases at or below the soil line. These lesions are bordered with a purplish-brown margin and may encircle 40-60% of the lower stem.

A premature nut drop of coconut in southeast Asia is also attributed to *P. palmivora*. Infection begins at the floral parts or the equatorial portion of the nut. The husk surface develops mottled areas with light brown centers and yellow margins. The fungus advances toward the apical end of the nut and produces irregular, brownish-black, oily husk lesions. Nuts less than 4 months old are most susceptible whereas those 10 months old are normally unaffected. This infection is often followed by secondary fungi like *Botryodiplodia, Chalara, Colletotrichum*, and *Fusarium* spp.

Fruit rot and mortality of coconut have been attributed also to *P. katsurae*. The initial disease symptom on mature trees is a nut rot

Cocos nucifera infected with *Phytophthora katsurae*.

Phytophthora bud rot of *Cocos nucifera* following a hurricane. (Courtesy Fl. Div. of Plant Ind.)

Phytophthora bud rot on mature *Cocos nucifera*.

Bud rot on palm caused by *Phytophthora palmivora*.

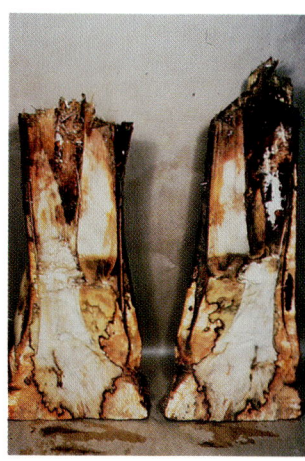

Heart rot on *Cocos nucifera* infected with *Phytophthora katsurae*.

Cross-section of *Carpentaria* stem infected with *Phytophthora*.

Phytophthora bud rot on field grown *Washingtonia* palms.

Above ground symptoms of Phytophthora root rot of *Howea forsterana*.

and drop of immature nuts in or under the sepals. Infected fruits have brown, irregular lesions which may be angular and elongate, or form large, mottled patterns surrounding circular green tissue areas. Lesion edges are frequently irregularly water-soaked. In contrast, healthy mature fruits, fruits damaged by insects, or fruits which abscise prematurely due to nutritional problems, are generally uniformly brown and not irregularly mottled. Internally, the husk and endosperm of Phytophthora-infected young fruits are dark and irregularly necrotic. Older mature fruits have dark brown to reddish brown internal husk tissue, while the endosperm is slightly discolored to very light brown. Secondary invasion by *Thielaviopsis paradoxa* will cause blackening of infected husks.

Several leaf spots and blight are caused by *P. palmivora* on *Chamaedorea elegans*. Leaf spots are brown to dark brown with tan centers and begin as slightly grayish-green irregular, circular to elongate, to very long lesions, which rapidly coalesce into blights. These blighted areas are dark brown to papery-tan with a dark border, somewhat vein limited by the midrib, but less so than on *Chrysalidocarpus* infected with *P. nicotianae*. Larger plants will exhibit a bud-rot phase before succumbing. On young seedlings of *Chamaedorea elegans* and *Chrysalidocarpus lutescens*, leaf lesions expand into the petiole and stem, killing the plant.

The occurrence of root rot in palms caused by *Phytophthora* is apparently infrequent. In a recent report, necrotic roots in declining bamboo palm (*C. seifrizii* x *erumpens*) were attributed to *P. arecae*, although the authors allude to the possibility that this fungus may be synonymous with *P. palmivora*.

Causal Organism: The most common species of *Phytophthora* causing diseases of palms is *P. palmivora* (Synonyms = *Phytophthora faberi* and *Phytophthora theobromae*). *Phytophthora palmivora* is a member of the Oomycetes. Hyphae of *P. palmivora* are typically narrow in diameter (5 μm) with few crosswalls except where reproductive structures are delimited.

Phytophthora palmivora produces large masses of deciduous sporangia on agar culture (10% V8 agar) under fluorescent irradiation near 24°C. Sporangia are papillate, ovoid, limoniform to ellipsoid, averaging 30 μm in diameter and 50 μm in length, with length to diameter ratios of 1.6 and higher, have short pedicels (<5 μm) and are borne sympodially on simple sporangiophores. Sporangia germinate by germ tubes or release zoospores in water. Chlamydospores are readily produced in large numbers and the species is heterothallic.

Phytophthora nicotianae readily produces non-deciduous sporangia on agar culture flooded with water. Sporangia are papillate, nearly spherical, and have a length to diameter ratio of less than 1.3, with an average diameter of approximately 34 μm. Chlamydospores are formed in this heterothallic species.

Phytophthora katsurae forms papillate sporangia with bilateral symmetry, approximately 24 μm wide. Chlamydospores have not been observed in this homothallic species. Oogonia have funnel-shaped tapered bases with amphigynous antheridia, and the oogonial walls are ornamented with wart-like or bulliform protuberances. These ornamentations are frequently absent or poorly developed in host tissue but are well developed in agar culture. The bulliform protuberances on oogonial walls and funnel-shaped tapered oogonial bases comprise a unique combination and allows ready recognition of *P. katsurae*.

Occurrence: *Phytophthora palmivora* is distributed worldwide in tropical and warm temperate regions that are characterized by high rainfall. Nations reporting this pathogen on palms include Ceylon, Columbia, Fiji, French Polynesia, Guatemala, Honduras, India, Indonesia, Ivory Coast, Jamaica, Nicaragua, Panama, Papua New Guinea, Philippines, Puerto Rico, Santo Domingo, Solomon Islands, Sri Lanka, Trinidad, United States (California and Florida), and Venezuela.

Closeup of root rot caused by *Phytophthora palmivora*.

Phytophthora nicotianae parasitizes many ornamental plants particularly in the tropics or sub-tropics. Several host range studies suggest that this fungus is not specialized, although a comprehensive study with a large collection of isolates is needed. Pending such a study, plants known to be infected with *P. nicotianae* should be kept away from palm crops.

Species Affected: Palms infected by *P. palmivora* are listed in Appendix 7.

Diagnostic Techniques: Diseased tissue can be microscopically examined for the presence of the mycelium of this pathogen. The hyphal strands have few crosswalls and ramify between cells with small, finger-like hyphal pegs extending into cells. Resolution of this mycelium will be enhanced by use of a cotton blue or aniline blue staining procedure. Chlamydospores may also be evident in diseased tissue. Incubation of symptomatic tissue in a high humidity environment for 24-48 hrs will stimulate sporangium formation. In general, isolation of *Phytophthora* is not very difficult, although some attention to the process is necessary. Specimens should be washed well in running water, using mild soap for waxy or hairy leaves, for uneven tissue surfaces, or for tissue heavily contaminated with secondary fungal growth. Interphase tissue between healthy and diseased areas of selected lesions should be excised. Avoid thoroughly necrotic tissue or the oldest parts of lesions, although these areas are ideal for detection of chlamydospores and oospores. Wash selected cut specimens in a sieve under running water for at least 10 min or dip specimens momentarily in 0.05-0.15% sodium hypochlorite. Blot on clean tissue paper and place on 1.7% water agar. Microbial growth on water agar is sparse, but mycelia of *Phytophthora* are easily recognized and sporangia usually form in 2-3 days. Hyphal tips or sporangia can be transferred to vegetable juice agar (V8 agar, carrot agar, etc.) preferrably in the first 1-2 days.

Roots infected with *Phytophthora palmivora*.

Phytophthora palmivora leaf spot on *Chamaedorea elegans*.

Internal bud tissues from the advancing margin of decay are best for isolation. External, decayed leaf base tissues are too frequently overrun with secondary invaders to ensure isolation of *P. palmivora*. Isolations also can be performed on a corn meal agar amended with antibiotics (recipes given in Appendix 8). The fungus in corn meal agar exhibits uniform growth, a radiate habit, little aerial mycelium and a temperature optimum of 27-32°C with a range of 11-35°C.

Phytophthora katsurae is difficult to isolate from badly diseased, rancid tissue in advanced heart rot of coconut. Microscopic

Phytophthora palmivora leaf spot on *Chamaedorea elegans*.

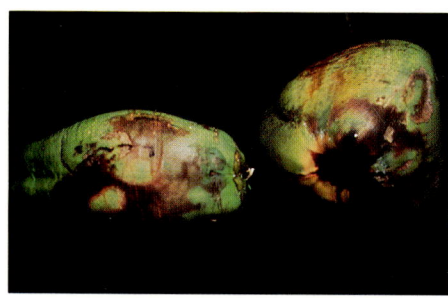

Cocos nucifera nut rot caused by *Phytophthora katsurae*.

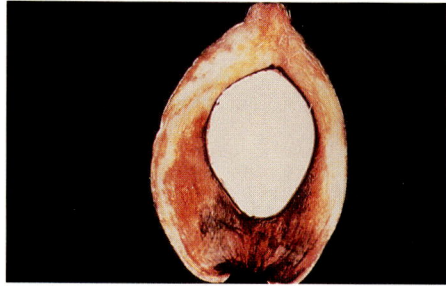

Cocos nucifera nut rot caused by *Phytophthora katsurae*.

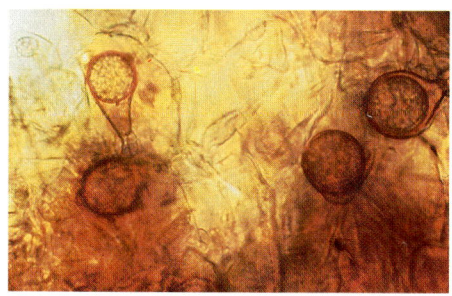

Phytophthora katsurae oospore.

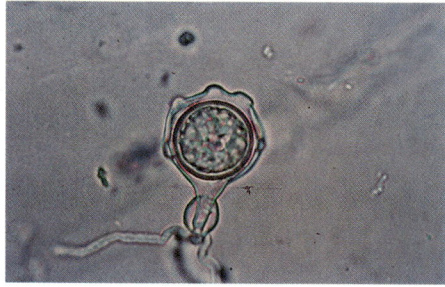

Phytophthora katsurae oospore.

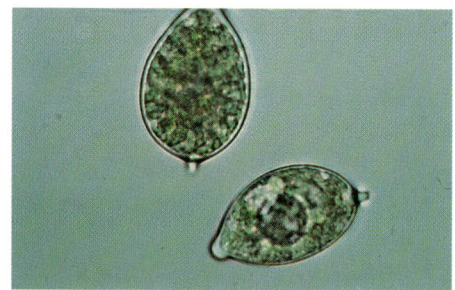

Sporangia of *Phytophthora palmivora*.

examination of the large necrotic heart reveals that *Phytophthora* is present at the leading edge of the lesion but only in certain areas. Much of the necrotic tissue is filled with bacteria and yeasts. In contrast, young fruit and leaf spot tissue are predominated by *Phytophthora*, thus, the pathogen is easy to isolate. This emphasizes the importance of tissue selection in successful isolation.

Phytophthora spp. are quite sensitive to hypochlorite solutions, thus, low concentrations and very short exposures should be used in disinfestations. Thin pieces of tissue which are readily penetrated by hypochlorite can be safely processed by extended washing in running water followed by plating on water agar.

Prevention and Treatment: Disease prevention cannot be overemphasized. Death is the inescapable result of heart rot of the sole apical meristem of solitary palms. Clean plant or seed sources are vital for disease avoidance and seeds should not be gathered from the ground, especially if the epicarp has rotted away. Seeds from unknown sources should be surface disinfested before planting, then carefully monitored for symptoms of seedling blights after planting. Early removal and destruction of infected plants will reduce inoculum levels and decrease the incidence of new infections. As with most fungal diseases, moisture control is all important in crop management. Free water favors the development of *Phytophthora* in all stages of its life cycle, and also favors and accelerates the processes in the disease cycle.

When weather favorable for disease persists, preventive use of fungicide sprays into the bud may be beneficial. Once disease occurs, severely affected palms should be rogued from either the nursery or the landscape and burned or buried. This will reduce the likelihood of pathogen spread to adjacent plants. Curative efforts with foliar fungicides will be most effective in limiting disease incidence and spread. Such products as fosetyl aluminum, metalaxyl, and propamocarb are broadly labeled for use on woody ornamentals and are all systemic products. In addition, copper-based fungicides will also provide a protectant layer on leaf bases and around the bud. Check product labels for legal use on palms.

Attempts to excise infected coconut heart tissue, followed by contact fungicide treatment have been unsuccessful in the control of *Phytophthora*. Some success in disease control has been achieved with injection of metalaxyl and fosetyl-Al into the trunk.

PINK ROT
(Gliocladium Blight)

Symptomology: This disease is an invasive rot that can attack the bud tissues, petioles, leaf blades and trunks. Often attacking trees under stress, the pathogen is believed to enter via wounds or areas damaged by such causes as removal of leaves, sunburn and freezing. Initially, dark brown necrotic areas appear on stems of infected plants near the soil line or 0.7 to 1.0 m up the stem. On *Syagrus*, infections can occur on the trunk at any height. Spots are often associated with gummy exudates. Older fronds die prematurely, necrotic streaks appear from the rachis base and pinnae turn chlorotic. The pathogen readily produces dusty masses of orange to pink conidia, often in sporodochia, on this tissue. In severe infections, many stems die due to girdling, giving potted plants an open unsalable appearance. Removal of symptomatic fronds reveals stem infections which are dark brown and irregularly shaped, sometimes with chlorotic margins.

Causal Organism: The first report of Gliocladium blight of ornamental palms was made by Bliss in 1938, with the causal agent identified as *Penicillium vermoeseni* now called *Gliocladium vermoeseni*. Conidia are borne on sterigmata and are 4-6 x 3-4 μ and elliptical when mature. Individual conidia are colorless but appear salmon or rosy en masse.

Occurrence: The pathogen is distributed worldwide causing disease on palms in greenhouses and landscapes.

Species Affected: Over the past 51 years many genera and species of palms have been listed as susceptible to this pathogen. Bliss isolated the pathogen from *Howea* sp., *Phoenix canariensis*, *Washingtonia filifera*, and *Syagrus romanzoffiana* in California. The disease was also found on *Chamaedorea seifrizii* in California in 1975. Most species of *Chamaedorea* and *Chrysalidocarpus lutescens* are highly susceptible to *G. vermoeseni*. Other palms which are susceptible include *W. robusta*, *Phoenix dactylifera*, and *Archontophoenix cunninghamiana*.

Diagnostic Techniques: The disease is generally diagnosed by abundant sporulation that occurs on host tissue. The pathogen is easily cultured on any general nutrition medium such as potato dextrose agar and corn meal agar. Radial growth is optimal at 27°C with no growth at 33°C.

Prevention and Treatment: Since wounding facilitates infection, only completely dead leaves should be removed from palms with Gliocladium blight. Removal of yellow leaves should be performed only when temperatures are at least 30°C, preferably 33°C, to reduce chances of infection. Applications of benomyl or benomyl and mancozeb at either 7- or 14- day intervals gave excellent disease control during the summer months in Florida on Chamaedorea palms. In the landscape, maintain plant health to avoid wounds. Transplanted specimen palms may benefit from prophylactic sprays of fungicides and stress reducing measures. On *Syagrus romanzoffiana*, trunk surgery followed by protective fungicides is reputed to effect a cure.

Pink rot at base of *Archontophoenix alexandrae*. (Courtesy D. Shaw)

Advanced symptoms of pink rot on *Syagrus romanzoffiana*.

Pink rot on *Washingtonia filifera*.

Pink rot on *Washingtonia filifera*.

Lower frond death on *Chamaedorea* sp. caused by *Gliocladium vermoseni*.

PSEUDOCERCOSPORA AND CERCOSPORA LEAF SPOTS

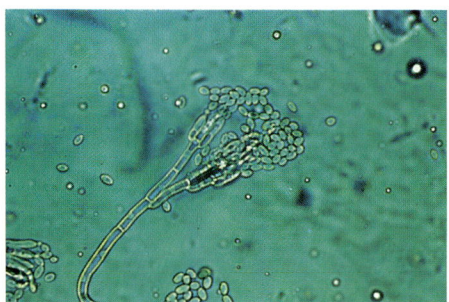

Massed conidia of *Gliocladium vermoseni* on lower fronds.

Conidiophore and conidia of *Gliocladium vermoseni*.

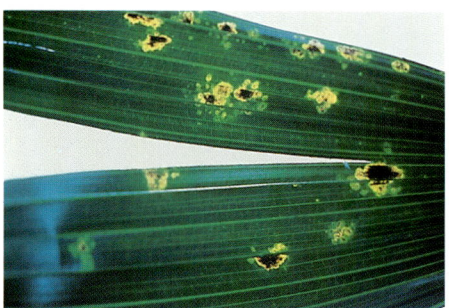

Pseudocercopsora leaf spot of *Rhapis excelsa*.

Pseudocercopsora leaf spot of *Rhapis excelsa* view with transmitted light.

Symptomology: Pseudocercospora leaf spot on *Rhapis excelsa* is a disease that develops very slowly. A period of 4 weeks or more may elapse between the time of infection and the appearance of initial symptoms. Leaf spots begin as very tiny, faintly chlorotic or light green flecks which are less than 0.5 mm in diameter. Over a period of 4-8 weeks, these expand into elliptical or circular spots, approximately 0.8-1.0 mm long, along parallel veins. Primary lesions expand, sporulate, and spores fall off to cause new infections of the tissue immediately adjacent to the primary lesions. This results in larger spots surrounded by pin point satellite lesions in the later stages of disease development. Leaf spots will eventually coalesce becoming irregularly circular to elliptical, averaging 6 by 12 mm. These areas are composed of slightly raised dark brown to reddish brown flecks and spots surrounded by chlorotic tissue which gradually turns brown. Light infections produce a few groups of spots while heavy infections produce leaves with mosaic patterns. Heavy infections result in premature leaf loss.

A leaf spot of *Elaeis guineensis* caused by *Cercospora elaidis* has been reported from Africa. The disease begins as tiny chlorotic spots, rarely larger than 0.5 mm in diameter, which eventually become brown and slightly sunken. Coalescence and heavy infections cause surrounding tissue to turn orange and become desiccated. The fungus grows very slowly on several culture media, producing septate, brown conidiophores and slightly olivaccous to pale brown conidia. The conidia are obclavate, with thickened basal scars, 3-9 septate and average 120 x 7 μm. A minor disease of *Licuala ramsayi* caused by a *Cercospora* species has been reported from Australia. Lesions are pin-point to 4 mm in size, gray to brown in color, with dark brown borders and yellow halos.

Causal Organism: This disease is associated with *Pseudocercospora rhapisicola* (=*Cercospora rhapisicola*) which was first reported from Japan. *Pseudocercospora rhapisicola* is transferred from *Cercospora* because of its unthickened conidial hilum and inconspicuous scars or conidiophores which are consistent with the description of *Pseudocercospora*. *Pseudocercospora rhapisicola* is readily isolated from leaf spots on rhapis palm. In culture, *P. rhapisicola* produces olive-green to gray-black colonies depending on the nutrient medium. Growth is slow but axenic culture is not difficult. Conidiophores are light brown to brown, with zero to few septa, rarely branched, with inconspicuous conidial attachment scars. Conidia are narrowly obclavate, straight or slightly curved, and have a tapered or conical base with unthickened hilum. Spores of a typical isolate measure 120.9±21.5 by 2.8±0.2 μm and have 12.9±4.1 septa; approximately 13% have a lateral appendage, up to 20 μm long, near the spore apex or base. No perithecial state has been observed.

Occurrence: *Pseudocercospora* is a notoriously slow grower and the diseases it incites are also very slow to develop. The long incubation period makes it extremely difficult to ascertain whether symptomless leaves are free from infection. The disease has been reported from Japan and in the United States (Hawaii, California and Florida). In Japan, the disease was found on *R. excelsa* but not on *R. humilis*. Other *Pseudocercospora* has been isolated from *R. subtilis*, *Howea* and *Chrysalidocarpus*.

Diagnostic Techniques: Slow disease development and the viral-like mosaic pattern of leaf spots on rhapis palm are important clues to this disease. Although *Pseudocercospora* grows very slowly, its isolation can be accomplished without undue difficulty. Leaf lesions should be rubbed with liquid hand soap, followed quickly by thorough rinsing under running tap water. This procedure greatly reduces or eliminates rapid-growing saprophytic contaminants which may overrun isolation plates. Well washed leaves should be placed on a clean paper towel or tissue paper, wiped, sectioned, surface sterilized in 0.1% hypochlorite solution for a few seconds, blotted momentarily and placed on the surface of 1.7% water agar. Following incubation of tissue sections at 20-25°C with light, conidia are produced after a few days, permitting the identification of *Pseudocercospora*. Similar methods can be used for *Cercospora*, which is slightly easier to culture. Cultural procedures for good growth and sporulation of *Pseudocercospora* and *Cercospora* are temperatures of 24°C or less; presence of light; and transfer of conidial inoculum to start new cultures to perpetuate sporulating capacity.

Prevention and Treatment: Close adherence to a program of sanitation and eradication of infected leaves is the key to *Pseudocercospora* disease control. Complete removal of diseased leaves followed by regular inspection and removal of new diseased leaves will reduce inoculum levels. By maintaining sanitation and keeping plants under solid-cover greenhouses, chemical treatment should be unnecessary. Applications of a fungicide such as maneb or mancozeb will reduce disease levels but only as a supplementary treatment to a sanitation program.

PSEUDOMONAS BLIGHT

Symptomology: The first symptoms of this disease are small water-soaked, translucent areas running along leaf veins. Mature lesions are brown to black, have a chlorotic halo, and range from 1-2 mm wide and up to 50 mm long. In many cases, the initial infection appears to occur at leaf margins through hydathodes.

Causal Organism: At the time Pseudomonas blight was described, the causal organism was identified as *Pseudomonas albo-precipitans* which more recently has been renamed *P. avenae*.

Occurrence: This disease has been reported on fish-tail palm in Florida only once in the past 20 years.

Species Affected: *Caryota mitis* is the only reported palm host of *Pseudomonas avenae*.

Diagnostic Techniques: Lesions should be excised from the advancing margins where water-soaked tissue is present. They can be crushed in sterile water, phosphate buffer or 0.01 M MgSO$_4$. The resulting suspension should be streaked onto standard nutrient media such as nutrient agar and incubated at 27-30°C for 2-4 days until

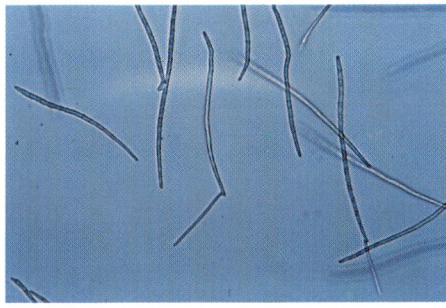

Conidia of *Pseudocercospora rhapisicola* with appendages.

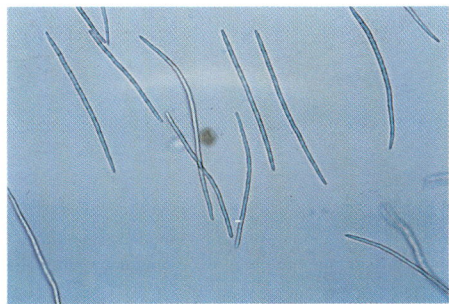

Conidia of *Pseudocercospora rhapisicola* without appendages.

Pseudomonas leaf spot of *Caryota mitis*.

Rachis blight of *Washingtonia filifera* caused by *Serenomyces california*.

Closeup of *Serenomyces california* on *Washingtonia filifera*. (Courtesy M. Murphy)

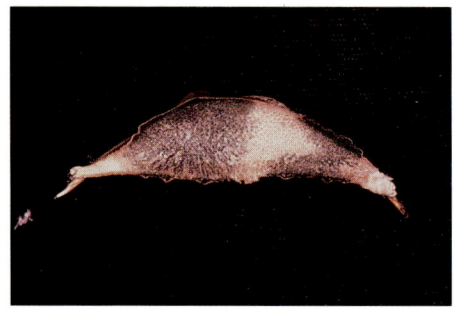

Cross-sections of rachis of *Washingtonia filifera* infected with *Serenomyces california*. (Courtesy M. Murphy)

Spores of *Serenomyces california*. (Courtesy M. Murphy)

Sclerotinia blight of *Chrysalidocarpus lutescens*.

discrete colonies appear. Successive transfer of single colonies should be made to prepare an axenic culture.

Strains of *P. avenae* from fish-tail palm are nonfluorescent, oxidase positive, reduce nitrate, produce a clear halo within a white precipitate when grown on lactalysate medium, and are negative for arginine dihydrolase production.

Prevention and Treatment: Initial water-soaked lesions appear in approximately 5 days. Typical symptoms of disease develop 7-10 days after infection when temperatures are between 18 and 32°C. No chemical controls have been investigated for this disease although copper or antibiotic products may be somewhat effective. Eliminating overhead irrigation and removing symptomatic leaves or entire plants are the recommended cultural controls.

RACHIS BLIGHT

Symptomology: The disease is characterized by the premature death of fronds and the appearance of abundant cinnamon-colored spores on the leaf rachis. Large necrotic, diamond-shaped areas appear on the rachis which, in cross sections, are found to extend through the interior thickness of the rachis and are tan to dark gray or black. Stromatic regions are dotted with small, rounded apical papillae surrounded with discharged ascospores which may collect into large cirrhi (up to 500 µm wide).

Causal Organism: *Serenomyces californica* is the cause of rachis blight of *Washingtonia filifera*. The asci, which are borne in locules, are thin-walled and about 40 x 25-30 µm. Ascospores are bright reddish-brown, unicellular, fusoid with short apical tips, 18-20 x 7-8 µm. Ascospores germinate readily on potato dextrose agar producing a fluffy white colony with optimum growth on vegetable juice agar at 30°C.

A second species, *S. phoenicis* has been reported from rachides of *Phoenix dactylifera*. Stromata of *S. phoenicis* are small and contain one or two stromata. Ascospores are 10-15 x 5-6 µm and have fine longitudinal striations.

Serenomyces sheari originally recovered from leaves of *Serenoa repens* was reported from Florida. Ascostromata are spheroid and 385-495 µm in diameter. Asci are 40 x 15-20 µm and ascospores are 13-17 x 5-6 µm with very slight longitudinal striations.

Finally, *S. palmae* was reported from an unidentified palm in Venezuela.

Occurrence: Rachis blight caused by one or more species of *Serenomyces* has been reported from the United States (California and Florida) and Venezuela.

Species Affected: *Washingtonia filifera, Phoenix dactylifera,* and *Serrenoa repens* are hosts of *Serenomyces* spp.

Diagnostic Techniques: The occurrence of the cinnamon-colored spores is diagnostic for *S. californica*.

Prevention and Treatment: No treatments are known although removal of infected foliage and use of a protective fungicide may help.

SCLEROTINIA BLIGHT

Symptomology: Sclerotinia blight occurs on small palm seedlings up to 50 cm tall. Foliar blight is accompanied by gray to white mycelia which frequently cover overlapping pinnae of affected fronds. Individual lesions are irregular in shape, surrounded by a water-soaked band of tissue, eventually turning tan to gray with a dark brown border.

Causal Organism: *Sclerotinia homeocarpa* causes this disease.

Occurrence: This disease is only rarely seen on *Chrysalidocarpus lutescens* in Florida.

Species Affected: *C. lutescens* is the only reported palm host although *S. homeocarpa* isolates from *Cynodon x magenissii* (Tifgreen bermudagrass) and *C. lutescens* are cross pathogenic.

Diagnostic Techniques: Isolates of *Sclerotinia homeocarpa* from *C. lutescens* grow best on malt agar. Aerial mycelia extend to the top of the petri dish and small (less than 1 mm in diam) sclerotia are produced near the lid of the dish. Mycelia are white to gray and flake-like, black, sclerotial tissue being produced within 3 to 4 weeks. Conidia or other fruiting structures have not been observed.

Prevention and Treatment: Sclerotinia blight of *C. lutescens* is most severe on plants less than 0.5 m tall, since they are densely planted (up to 100 seeds in a 20-cm pot) creating ideal conditions for infection and disease development. Disease incidence is highest during March and April in south Florida. Excellent control of Sclerotinia blight of *C. lutescens* was achieved with applications of a combination of benomyl and mancozeb.

STIGMINA LEAF SPOT
(Exosporium Leaf Spot)

Symptomology: This foliar disease develops during the cool seasons in both hemispheres. Leaf spots develop first on lower, older leaves and appear as small, round, yellowish lesions. Mature lesions are tan to reddish-brown in color, 0.5-1.0 cm in diameter, with a dark brown to black, depressed lesion center, and are often surrounded by a diffuse yellow halo. Leaf spots may merge killing large irregular areas of tissue. Both leaf pinnae and rachides can be infected.

Causal Organism: *Stigmina palmivora* (Synonym = *Exosporium palmivorum*) is a dark-colored, imperfect fungus for which no sexual stage is known. This pathogen invades directly through either leaf surface. The fungus ramifies below the leaf epidermis and emerges mainly from the upper leaf surface during reproduction. Conidiophores are dark brown to black, ranging in size from 12-30 μm long (av. 21 μm) by 4-7 μm wide (av. 4-5.5 μm) and terminating in as many as 5 ring-like cells. The conidia are borne either terminally or laterally from the conidiophores and are brown, thick-walled, straight to slightly curved, and may appear either smooth to coarsely warted. Measurements of conidia range from 60-120 μm long (mostly 64-89 μm) by 7-10 μm wide (4-7 μm at the

Lesions on *Chrysalidocarpus lutescens* caused by *Sclerotinia homeocarpa*.

Stigmina leaf spot on lower fronds of *Phoenix roebelinii*.

Closeup of Stigmina leaf spot on of *Phoenix roebelinii*.

Stigmina palmivora conidia and conidiophores. (Courtesy R. Cullen)

Thielaviopsis bud rot on *Phoenix canariensis*.

Thielaviopsis trunk decay on *Phoenix canariensis* showing masses of black chlamydospores.

Thielaviopsis stem rot on small *Chamaedorea elegans*.

base and 3-6 μm at the rounded apex).

Occurrence: This pathogen occurs primarily throughout the tropical regions. Reports exist from Australia, Ghana, Kenya, New Guinea, Pakistan, Sierra Leone, Uganda, United States (Florida, Louisiana, Mississippi, and Texas), West Indies, and Zimbabwe.

Species Affected: Some hosts of *Stigmina* are given in Appendix 9.

Diagnostic Techniques: The upper leaf surface of fresh lesions can be sectioned and/or scraped and examined for conidia. Leaf tissue can be incubated in a high humidity chamber for 24-48 hr to induce greater sporulation. The pathogen can be cultured on general growth media like potato dextrose agar (pH = 5.5).

Prevention and Treatment: Severity of this disease is negligible except in the nursery. Landscape palms planted in areas with poor air circulation and partial shade may develop disease on lower fronds. Removal of infected fronds is quite effective. In nursery situations where overhead irrigation and partial shade conditions exist, disease severity on containerized palms may be high. In the southeastern United States, observations indicate that Stigmina leaf spot is most severe in late fall through spring. This period is typified by cool temperatures and frequent leaf wetness cycles. In addition to sanitation, use of fungicides may be beneficial. Fungicide efficacy trials with a related *Stigmina* sp. on stone fruits indicate that compounds such as benomyl, iprodione, and copper-based materials are useful in reducing disease severity.

THIELAVIOPSIS BUD ROT
(Stem Bleeding, Bitten Leaf, Black Scorch, Dry Basal Rot, Heart Rot)

Symptomology: The pathogen *Chalara (Thielaviopsis) paradoxa* causes a range of symptoms on palms. This fungus incites a bud rot and the associated 'bitten leaf' symptoms. Invasion occurs down along the spear leaf or through a young leaf base into the bud region. Blackish-brown lesions develop on external and internal tissues. Abundant chlamydospore production in the tissue confers the dark color. Buds die or are damaged. Lateral bud development may occur, bending the direction of new growth. In addition, new leaves may exhibit the 'bitten leaf' appearance as they emerge deformed, with reduced pinnae, and black, necrotic tips.

Two other disease syndromes may occur in the palm crown. The fungus invades young emerging fronds causing a 'black scorch'. Dark-brown to black irregular lesions develop along the petiole and confer a torched appearance to tissue. Pinnae distort and/or shorten during emergence. Similarly, the inflorescence may be invaded prior to spathe opening. Round to elongate lesions, sorghum-brown to mahogany-brown in color, develop on the spathe. Young fruit stalks can exhibit similar lesions while the fruit strands and flowers may be partly or totally blackened. The trunk and roots are directly invaded by this pathogen. Mechanical damage or growth cracks from irregular moisture levels may provide entrance of this fungus into the trunk. A soft, yellow decay of the trunk tissue develops, darkening to black with age. As decay progresses, a reddish-brown or rust-colored liquid bleeds from the point of invasion. This sap flow may extend several feet down the trunk, blackening the trunk as it dries. Several infections may merge causing a heart rot. Affected palms have reduced growth and a gradual but progressive necrosis of lower leaf pinnae toward the midribs. Palms defoliate and die. The trunk is hollow due to the decay of the interior tissues. Roots may be similarly decayed. This pathogen causes several other diseases of palms outside of the western hemisphere.

Causal Organism: *Ceratocystis paradoxa* (sexual state) (Synonyms = *Ceratostomella paradoxa*, *Endoconidiophora paradoxa*) with *Chalara paradoxa* (Synonym = *Thielaviopsis paradoxa*) as the asexual state is the cause of this disease. The sexual state is a perithecium that is mostly immersed in the growth medium. It is pale to dark brown in color, globose in shape, measuring 190-260 μm in diameter, and ornamented with simple to forked, knobbed appendages. The neck of the perithecium is pale brown to black, measuring 700-1500 μm long with associated hyphal filaments at the opening. The ascospores are colorless, enclosed in a gelatinous sheath, measure 5.5-9.5 x 2-6.5 μm, and appear elliptical to boat-shaped. These spores are released in a mucilaginous chain.

The *Chalara* state (asexual) produces both endo conidia and chlamydospores. The conidiophores are usually straight, colorless to pale brown, up to 250 μm long, with a terminal spore-bearing cell through which spores are borne. The conidia are cylindrical to elliptical with square ends, colorless to pale brown, and measure 7-15 x 2.5-6 μm. The chlamydospores are borne terminally in chains from short hyphal branches, are pale brown to brownish black, smooth, oval, and measure 9.5-25 x 5.5-15 μm.

Ceratocystis radiciciola (Synonym = *Ceratostomella radicicola*) has the asexual state *Chalara radicicola* (Synonym = *Chalaropsis radicicola*). The sexual state is a perithecium that is wholly or partially immersed in the medium surface. This structure is pale brown in color, about 300 μm in diameter, with a variable number of irregularly tipped appendages measuring 85 x 8 μm. The necks of these flask-shaped perithecia are black, approximately 800 μm in length. Numerous, colorless filaments are arrayed about the opening of the neck and these measure 60 x 3 μm. Asco spores are elongate to ellipsoidal, often with one flat side and one tapered end and measure 7-12 x 2.5-3.5 μm in size.

Chalara radicicola has two spore forms like *C. paradoxa*. The conidiophore is colorless to pale brown, tapered in shape, septate, thin-walled, and measures 125-250 μm. The endo conidia are similarly colored, cylindrical to round in shape, thin-walled, mostly borne in chains, and measure 6-22 x 4-5 μm. The chlamydospore is terminal, solitary, with a slightly ornate wall, and measures 9-24 x 7-15 μm.

Occurrence: *Chalara (Thielaviopsis) paradoxa* is reported from Algeria, Brazil, Cameroon, Columbia, Dominican Republic, Ecuador, Egypt, El Salvador, Gayana, Ghana, Iraq, Jamaica, Mauritania, Mexico, Nigeria, Philippines, Puerto Rico, Saudi Arabia, Sri Lanka, Trinidad, Tobago, Tunisia, United States (Arizona, California, and Florida), Venezuela, and the West Indies. *Chalara*

Closeup of Thielaviopsis stem rot on small *Chamaedorea elegans*.

Thielaviopsis leaf spot on *Chamaedorea elegans*.

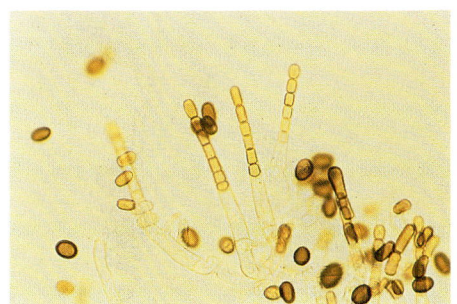

Thielaviopsis paradoxa conidia and chlamydospores. (Courtesy R. Cullen)

Albino seedlings of *Caryota mitis*.

B-deficient *Chamaedorea elegans*.

B-deficient *Chrysalidocarpus lutescens*.

B-deficient *Caryota mitis* seedling.

radicicola is reported from California.

Species Affected: *Areca catechu, Brahea edulis, Caryota* spp., *Cocos nucifera, Elaeis guineensis, Phoenix africanus, Phoenix canariensis, Phoenix dactylifera, Rhapis* sp., *Roystonea elata, Sabal palmetto, Syagrus romanzoffiana*, and *Washingtonia filifera* are reported as hosts of *C. paradoxa*. In addition, *Chalara paradoxa* has a wide host range outside the Arecaceae which includes many economically important hosts. Rhizosis disease caused by *Chalara radicicola* is reported from only *Phoenix* spp.

Diagnostic Techniques: Isolation of *Chalara paradoxa* from infected palm tissues is easy on a broad array of general growth media. This pathogen grows well between 25-32°C on media with a pH=3.6. Mycelial growth on potato dextrose agar at 30°C can exceed 45-60 mm in 5 days. The initial culture is hyaline to white in color, developing abundant aerial mycelium. As the culture ages, mycelial color will shift to gray and then to greenish-black within 10 days. This species has no characteristic odor in culture. Glucose and asparagine are the optimal carbon and nitrogen sources for growth, respectively. Conidial production usually precedes chlamydospore production by a period of hours although both states may begin within 24 hrs of colony development. Chlamydospores are abundant, smooth-walled, and borne in chains in culture. The perfect stage is rare.

Ceratocystis (Chalara) radicicola can be found throughout necrotic cortex and stele tissues of palm roots. The chlamydospore stage is quite abundant in these tissues. This species cultures readily on general growth media, growing more dense on potato dextrose agar than on corn meal agar. Young mycelium is hyaline and fast growing. Mycelia will develop a greenish-gray color after a few days and cultures have a diagnostic odor of banana oil. Chlamydospores have ornamented walls and are borne singly and terminally. The perfect state is rare in culture.

Prevention and Treatment: Avoidance of wounds in palms grown in field nurseries or in the landscape can limit disease incidence. Infected plants should be cut, removed, and destroyed. In some cases, localized infections can be excised through tree surgery followed by benomyl treatment. Although some resistance to this pathogen has been documented in date palm, the range of disease reactions is not known for ornamental palm species. Early infection of bud, leaf, or root tissue can be effectively treated by benomyl. This fungicide has also proven effective as a oil palm seed treatment to prevent black rot on seedlings.

PHYSIOLOGICAL DISORDERS

ALBINISM

Symptomology: Seedlings emerge white and totally devoid of chlorophyll. Tips of leaflets become necrotic quickly and the seedlings soon die.

Cause: Albinism is genetically controlled and seeds from some palms may produce mostly albino seedlings. Occasional albino seedlings may also occur among normal seedlings.

Occurrence: Albinism is rather uncommon in palms.

Species Affected: Albinism can affect any palm species.

Diagnostic Techniques: Visual symptoms are sufficient to identify this problem.

Prevention and Treatment: Do not use seed from palms known to produce albino seedlings.

BORON DEFICIENCY

Symptomology: Early boron (B) deficiency symptoms on *Chamaedorea elegans*, *Caryota mitis*, and *Phoenix roebelenii* include chlorotic new leaves that are usually malformed and fail to expand normally. Growth rate is greatly reduced in affected palms. Leaf margins are often necrotic and in more severe cases, entire leaflets may be necrotic. In the most severe cases, only necrotic petiole stubs will emerge and death of the meristem will follow.

In *Cocos nucifera* and *Elaeis guineensis*, B deficiency is often called "Hookleaf" due to the necrosis and subsequent withering of the leaflet tips. More severe symptoms are similar to those of *Chamaedorea* and *Phoenix*. Flowers often blacken and abort, and fruit may not set or will drop in B-deficient palms.

Cause: B deficiency is usually caused by insufficient B in the soil. B is readily leached from most soils.

Occurrence: B deficiency is rare in landscapes in the United States, but is common in some commercial coconut and oil palm growing regions of the world. B deficiency can occur in container-grown palms if no B fertilizers are used.

Species Affected: Most species of palms are probably susceptible to B deficiency.

Diagnostic Techniques: Leaf nutrient analysis, in addition to visual symptoms, is usually necessary to diagnose B deficiency. See Appendices 10-12 for critical B concentrations in some palms.

Prevention and Treatment: Soil or foliar applications of sodium borate or boric acid will treat B-deficient palms. B should not be applied indiscriminately since B toxicity can occur at fairly low levels.

B-toxicity on older leaf of *Chamaedorea elegans*.

New leaves of Ca-deficient *Howea forsterana*.

Ca-deficient *Chamaedorea elegans*.

Cl-deficient *Caryota mitis* seedling.

Cl-deficient *Phoenix roebelinii* seedling.

Cold damage on *Geonoma* sp.

Freeze damage on *Chrysalidocarpus lutescens*.

Freeze damage on *Cocos nucifera*.

BORON TOXICITY

Symptomology: Older leaflets of affected palms have light tan necrotic tips, the necrosis gradually spreading up the leaflets.

Cause: This disorder is caused by excessive B in the soil or irrigation water.

Occurrence: B toxicity is rather rare in areas of high rainfall, but can occur in drier climates and in container-grown palms if too much B fertilizer is applied or high B irrigation water is used.

Species Affected: B toxicity has been shown only on *Chamaedorea elegans*, but many other species are probably susceptible.

Diagnostic Techniques: Leaf B analysis along with visual symptoms are needed to diagnose this disorder.

Prevention and Treatment: Avoid using water high in B or excessive rates of fertilizer with high B. Palms suffering from B toxicity can be treated by heavy soil leaching.

CALCIUM DEFICIENCY

Symptomology: Symptoms of young calcium (Ca) deficient *Chamaedorea elegans*, *Howea forsterana*, and *Elaeis guineensis* include stunted, deformed new leaves that fail to expand normally. Leaflets of newly emerging leaves become necrotic with only the petiole base remaining alive. In succeeding leaves, only necrotic petiole stubs emerge, the leaflets and most of the rachis having died before completely developing. The necrotic petiole stubs are usually watersoaked in appearance. Severe Ca deficiency will kill the meristem.

Cause: Ca deficiency is caused by insufficient Ca in the soil.

Occurrence: Ca deficiency is rare in field soils, but can occur in container media if no dolomite has been added.

Species Affected: Most species of palms are probably susceptible to Ca deficiency.

Diagnostic Techniques: Since visual symptoms are somewhat similar to those of B, Zn, and Mn deficiencies, leaf nutrient analysis should also be used. See Appendices 10-12 for critical Ca concentrations in some palm species.

Prevention and Treatment: Ca deficiency is best prevented by amending all container media with dolomite. Palms already deficient in Ca may respond more quickly to foliar sprays of calcium nitrate than to soil applications of Ca fertilizers, but sustained growth will only occur if Ca sources are also applied to the soil.

CHLORINE DEFICIENCY

Symptomology: Chlorine (Cl) deficiency symptoms in *Caryota mitis* consist of mildly chlorotic new leaves, but in *Phoenix roebelenii*, chlorosis is more severe and is accompanied by incomplete separation of new leaflets. These leaves have a ladder-like appearance, with leaflets being attached at the margins so securely that they cannot be separated without tearing the leaflets. Fruit size and production, and plant vigor are reduced in Cl deficient *Cocos nucifera* and *Elaeis guineensis*.

Cause: Cl deficiency is caused by insufficient Cl in the soil. This ion is readily leached from most soils, but is usually replaced by Cl from the atmosphere.

Occurrence: Cl deficiency has been documented in *Cocos* and *Elaeis* plantations in the Philippines, Ivory Coast, Peru, and Indonesia, but is not known to occur in the United States.

Species Affected: Cl deficiency occurs naturally in *Cocos* and *Elaeis* and has been induced experimentally in *Caryota* and *Phoenix*.

Diagnostic Techniques: Although visual symptoms in *Phoenix* are quite distinctive, symptoms in other palms are not and leaf analysis may be required for accurate diagnosis. Cl is required in only micronutrient quantities (10-30 ppm) by most palms, but is usually present in macronutrient concentrations (0.2-2.0%) in the foliage. In *Cocos*, leaves with Cl concentrations of 0.3% or less are considered deficient, while 0.5% Cl is considered optimal.

Prevention and Treatment: Cl deficiency can be prevented and treated by applying KCl or NaCl to the soil. Care should be taken to apply accurate amounts since excess salts can cause root damage and eventual loss of the palm in some species. In most areas, Cl from air pollution, seaspray, or saline irrigation water provide adequate amounts of this element to replace that lost due to leaching.

COLD INJURY

Symptomology: Primary symptoms include extensive necrosis of the foliage. Necrosis is not necessarily confined to leaflet tips. Symptoms are very similar to those of water stress and often include wilting. Unhardy palms may be killed. Secondary symptoms include Erwinia bud rot and transient micronutrient deficiency symptoms.

Cause: Exposure to temperatures below that to which palms are acclimated will cause injury. Palms differ greatly with respect to their cold tolerance, both within and between species. Within a species, susceptibility to cold injury is reduced by gradually acclimating palms to cooler temperatures. Temperatures of 10°C can injure many tropical palms, while some temperate species can withstand freezing temperatures with no injury. Palms stressed by cold temperatures are very susceptible to Erwinia bud rot and may later show signs of micronutrient deficiencies such as Mn due to lower root activity at cooler temperatures.

Species Affected: All species are susceptible, but the temperature at which injury will occur is dependent on the species.

Diagnostic Techniques: Symptoms are used for diagnosis.

Prevention and Treatment: Grow only palms that are adapted to your climate. Small, valuable palms can be covered during occasional freezes. Palms in containers can be moved indoors for protection during cold weather. Some recommendations suggest

Freeze damage on *Veitchia mcdanielsii*.

Mn-deficiency induced by low temperatures on *Cocos nucifera*.

Cu-deficient *Phoenix roebelinii*.

Cu-deficient *Howea forsterana* with tip necrosis on new leaf.

Cu-deficient *Howea forsterana*.

Cu-deficient *Chamaedorea elegans* (left).

Cu-toxicity on *Chrysalidocarpus lutescens*.

Cu-toxicity on *Butea capitata*.

drenching the bud area with a copper-containing fungicide to prevent bud rot. To prevent transient Mn deficiency, apply Mn sulfate in the fall to the soil. A foliar spray with manganese sulfate in the spring may help Mn-deficient palms recover more quickly.

COPPER DEFICIENCY

Symptomology: Symptoms of copper (Cu) deficiency in *Chamaedorea elegans*, and *Howea forsterana* include new leaves that are reduced in size and have necrotic margins. As the symptoms progress only necrotic-tipped petioles emerge and death of the meristem usually follows. Symptoms are similar to those of Mn and Zn deficiencies except that chlorosis is less prevalent in Cu-deficient *Chamaedorea* and non-existent in *Howea*. In *Phoenix roebelenii*, new leaves emerge chlorotic and malformed. New leaves of Cu-deficient *Elaeis* have chlorotic leaflet tips that eventually become necrotic from the leaf tip downwards.

Cause: Cu deficiency results from insufficient available Cu in the soil. Cu is very tightly bound by most soils, especially those with high organic matter content.

Occurrence: Cu deficiency has been reported from Hawaii, but this has not been documented. Cu deficiency is associated with the "mid-crown chlorosis" disorder of *Elaeis guineensis* grown in Malaysian peat soils.

Species Affected: Cu deficiency has been experimentally induced in *Chamaedorea, Phoenix,* and *Howea*, and occurs naturally in *Elaeis*, but other species may be susceptible as well. *Cocos nucifera* is apparently quite resistant to Cu deficiency.

Diagnostic Techniques: Since visual symptoms are similar to those of other micronutrient deficiencies, leaf analysis is required for accurate diagnosis of Cu deficiency. See Appendices 10-12 for critical Cu concentration for some palm species.

Prevention and Treatment: Cu deficiency can be prevented or treated with soil applications of Cu salts and short-term treatment with foliar sprays of neutral copper sulfate are also effective. Continued use of Cu-containing fertilizers or fungicides can result in toxicity however.

COPPER TOXICITY

Symptomology: Cu toxicity in palms is usually manifested as distinct 1-4 mm diameter necrotic spots on the leaflets, but in some species such as *Butia capitata* extensive tip and marginal necrosis are also typical symptoms.

Cause: Cu toxicity is usually caused by foliar applications of Cu-containing fungicides or fertilizers. Excessive soil applications of Cu fertilizers could also potentially result in Cu toxicity.

Occurrence: Cu toxicity frequently occurs when Cu-containing fungicides or fertilizers are applied to palm foliage, but Cu toxicity due to high Cu levels in the soil is much less common due to strong Cu binding by most soils.

Species Affected: Most palms are probably susceptible to Cu toxicity, but their relative tolerances to high Cu vary.

Diagnostic Techniques: Visual symptoms, plus a knowledge of Cu fertilizer or fungicide applications, are usually sufficient for diagnosis, but leaf elemental analysis is also helpful.

Prevention and Treatment: Avoid the use of Cu-containing fertilizers in foliar sprays and use Cu-based fungicides sparingly, if at all to prevent Cu toxicity. Usually affected palms will eventually grow out of the symptoms without treatment, but injured leaves will remain on the palm until they are replaced by new foliage.

EXCESSIVE WATER UPTAKE

Symptomology: Palms that take up excessive amounts of water may have trunks with deep longitudinal splits. These trunks will often appear waterlogged and will likely be covered with mosses and other epiphytes.

Cause: This splitting is caused by excessive water uptake.

Occurrence: This problem is rare in most palm-growing regions, but not unusual in areas such as parts of Hawaii that receive very high rainfall.

Species Affected: Trunk splitting due to excessive water uptake has been documented for *Archontophoenix alexandrae*, but probably occurs in other species as well.

Diagnostic Techniques: Visual symptoms should be sufficient to diagnose this disorder.

Prevention and Treatment: This problem cannot be treated or prevented.

FLUORIDE TOXICITY

Symptomology: Leaflets of affected palms have necrotic lesions, usually arranged in longitudinal rows.

Cause: This disorder is caused by toxic levels of fluoride (F) in the soil solution.

Occurrence: F toxicity is rare in palms, but could possibly occur when irrigation water containing high F is used.

Species Affected: *Chrysalidocarpus lutescens* is slightly susceptible to injury by high F in the water, but *Chamaedorea elegans* is very tolerant. Most palms are probably quite tolerant of F.

Diagnostic Techniques: Leaf F analysis is needed to confirm F toxicity, since symptoms are similar to those of some fungal leaf spot diseases.

Prevention and Treatment: Avoid using F-containing materials such as ordinary superphosphate fertilizer and perlite, or irrigation water containing high levels of F. Leach thoroughly to remove F from the soil.

Trunk of *Archontophoenix alexandrae* suffering from excess water uptake.

F-toxicity on *Chrysalidocarpus lutescens*.

Foliage of *Syagrus romanzoffiana* injured by salt spray.

Flowered *Metroxylon saga*.

FOLIAR SALT INJURY

Symptomology: Foliar salt injury appears as desiccation of the foliage. Symptoms may be more severe on the exposed windward side of palms growing near the ocean. Some palms can be killed by salt spray.

Cause: Salt spray from the ocean on the foliage can cause desiccation of leaves unless it is rinsed off by rain or other fresh water.

Occurrence: Foliar salt injury is fairly common along seashores during dry periods with strong onshore winds.

Species Affected: Many species are affected by salt spray, but *Cocos nucifera*, *Serenoa repens*, *Allagoptera arenaria*, *Hyophorbe* spp., *Coccothrinax* spp., *Thrinax* spp., and *Pseudophoenix* spp. are quite tolerant of seashore conditions.

Diagnostic Techniques: Visual symptoms are usually sufficient to diagnose this problem.

Prevention and Treatment: Plant only salt-tolerant species of palms in areas exposed to salt spray. If strong dry onshore winds persist, foliage of palms should be rinsed with fresh water soon after these conditions subside.

Caryota mitis declining after flowering.

HAPAXANTHIC FLOWERING

Symptomology: Shoots of flowering or fruiting palms gradually decline and die.

Cause: In hapaxanthic palm species, shoots normally die after flowering and fruiting.

Occurrence: Death due to flowering and fruiting is common in affected species.

Species Affected: *Caryota* spp., *Arenga* spp., *Corypha* spp., *Wallichia* spp., *Nannorrhops ritchiana*, *Metroxylon* spp., and *Raphia* spp. are hapaxanthic.

Diagnostic Techniques: Visual symptoms are sufficient for diagnosis.

Prevention and Treatment: Plants only pleonanthic palms for long-term landscape use.

Flowered *Corypha elata*.

HERBICIDE TOXICITY

Symptomology: Symptoms of herbicide toxicity vary with the compound. Glyphosate injury appears as distortion and reduction in size of new leaves. New leaflets may show some necrosis. 2,4-D injury appears as distortion of the foliage. Pre-emergent herbicides containing oxyfluorfen, pendimethalin, or oryzalin result in new leaves having necrotic blotches. Paraquat toxicity appears as desiccation of the foliage contacted by the herbicide. Mature, established palms are usually not killed by any of the above herbicides.

Cause: Many herbicides can cause injury to palms.

Occurrence: Herbicide toxicity is fairly common.

Pritchardia sp. injured by the herbicide 2,4-D.

Ptychosperma macathurii injured by the herbicide glyphosate.

Chamaedorea elegans injured by oxyfluorfen and pendimethalin herbicides.

Washingtonia robusta injured by tebuthuiron herbicide.

Sabal palmetto injured by brackish water.

New leaf of Fe-deficient *Rhapis excelsa*.

Glyphosate injury on young *Phoenix roebelinii*.

Old leaf of *Syagrus romanzoffiana* showing injury due to high soluble salts.

Lodging of *Neodypsis* sp. caused by excess fertilizer.

New leaf of Fe-deficient *Chamaedorea elegans*.

Old leaf (left) and new leaf (right) of Fe-deficient *Phoenix roebelinii*.

New leaf of Fe-deficient *Caryota mitis*.

New leaflets of Fe-deficient *Syagrus romanzoffiana*.

Lightning injury on *Syagrus romanzoffiana*.

Species Affected: *Cocos nucifera* is highly tolerant of glyphosate applied to the foliage, while most other species of palms are slightly to moderately injured by direct sprays with this material. Seedlings of most palms are injured when pre-emergent herbicides containing oxyfluorfen, oryzalin, or pendamethalin are used, even at recommended rates. No palms are known to be injured by oxadiazon when it is applied as recommended. The relative susceptibility of palms to other herbicides is not known.

Diagnostic Techniques: Visual symptoms and a knowledge of treatment history are needed to diagnose herbicide toxicity.

Prevention and Treatment: Do not allow any herbicide to contact palm foliage. Glyphosate and paraquat are very safe around palms if kept off the foliage. If a pre-emergent herbicide is needed, use only oxadiazon. Palms exhibiting signs of herbicide toxicity will usually grow out of the symptoms in a few months, even if nothing is done.

HIGH SOIL SOLUBLE SALTS

Symptomology: Palms suffering from high soil soluble salts usually have necrotic leaflet tips on older leaves. New foliage may emerge chlorotic and growth rate is reduced. Roots will often have necrotic tips or more extensive necrosis.

Cause: High soil soluble salts can be caused by excessive fertilization or the use of high salt fertilizers, or by using saline irrigation water.

Occurrence: Soluble salts injury can occur under landscape conditions, particularly in drier climates and in container production of palms.

Species Affected: Most species can be affected, but salt-tolerant species such as *Cocos nucifera*, *Phoenix* spp., *Hyophorbe* spp., *Thrinax* spp., *Coccothrinax* spp., *Allagoptera arenaria*, and *Serenoa repens* are less likely to be injured.

Diagnostic Techniques: Visual symptoms plus soil soluble salts analysis are needed to diagnose this disorder.

Prevention and Treatment: Low salt fertilizers should be used and only at recommended rates. Some leaching of the soil should occur at each irrigation to prevent the problem. If the problem is caused by saline irrigation water and a cleaner water source cannot be found, the soil should not be allowed to dry out and/or only salt-tolerant species of palms should be grown. If the condition already exists, the soil should be leached thoroughly several times to remove excess salts.

IRON DEFICIENCY

Symptomology: Iron (Fe) deficiency appears as interveinal or uniform chlorosis or chlorotic spotting of the newest leaves. Older leaves remain green. In severely Fe-deficient palms, new leaflets may have necrotic tips, growth will be stunted and the meristem may eventually die. Early symptoms of Fe deficiency in *Syagrus* *romanzoffiana* appear as chlorotic new leaves covered with green spots about 1-2 mm in diameter.

Cause: Fe deficiency is not usually caused by a lack of Fe in the soil, but rather by poor soil aeration and by planting palms too deeply. Both factors reduce root uptake of iron. In alkaline soils, Fe may not be readily available to plants, but in palms Fe deficiencies are rarely caused by high pH.

Occurrence: Fe deficiency is rather uncommon, but can occur in containers or landscapes if the soil is poorly aerated or the palm is planted too deeply.

Species Affected: Most species of palms are susceptible to Fe deficiency.

Diagnostic Techniques: Visual symptoms are very similar to those of S deficiency and leaf nutrient analysis is needed to differentiate between the two disorders. See Appendices 10-12 for critical Fe concentrations for some palm species.

Prevention and Treatment: Fe deficiency is best prevented by planting palms no deeper than they were originally growing and by planting in a well-drained soil or site. Applications of Fe fertilizers to the soil are often ineffective in correcting the problem, but foliar sprays of ferrous sulfate or iron chelates may temporarily alleviate the symptoms. Long-term treatment must include correction of the cultural problem that induced the Fe deficiency in the first place. Chlorotic Fe-deficient foliage usually regains its normal green color following treatment, but necrotic tissue will remain.

LIGHTNING INJURY

Symptomology: Lightning injury symptoms are quite variable, but often include dark longitudinal streaks, bleeding, and splitting of the trunk, and sudden collapse of the crown. Lightning strikes often kill palms.

Cause: This disorder is caused by lightning striking the palm.

Occurrence: Lightning injury is fairly common in areas receiving frequent electrical storms.

Species Affected: Any species is susceptible to lightning injury, but usually only taller species will be struck.

Diagnostic Techniques: Visual symptoms are sufficient for diagnosis.

Prevention and Treatment: Lightning injury cannot be treated and prevention by lightning grounds is impractical for all but the most valuable palms.

MAGNESIUM DEFICIENCY

Symptomology: Oldest leaves of magnesium (Mg)-deficient pinnate-or costapalmate-leaved palms usually have broad chlorotic bands along the margins with the chlorosis starting at the leaflet tips and expanding towards the rachis as the deficiency progresses. In severe cases, only the rachis and adjacent portions of the leaflets remain green on the oldest leaves, but younger leaves show

Lightning injury on *Cocos nucifera*.

Older leaf of Mg-deficient *Dictyosperma album*.

Older leaf of Mg-deficient *Livistona rotundifolia*.

Mg-deficient *Pritchardia* sp.

Older leaf of Mg-deficient *Chamaedorea elegans*.

Mg-deficient *Phoenix canariensis*.

New foliage of Mn-deficient *Elaeis guineensis*.

New foliage of Mn-deficient *Roystonea regia*.

Mn-deficient *Syagrus romanzoffiana*.

progressively wider bands of green along the centers of leaves. In *Phoenix* spp., leaflet tips may become necrotic in severe cases. In palmate-leaved palms symptoms are similar except that the chlorosis appears as a broad yellow band around the margins of the oldest leaves, the center of the leaf remaining green. In species with deeply dissected leaves, the chlorosis appears as broad bands on the margins of each leaflet tip.

Cause: Mg deficiency is caused by insufficient Mg in the soil. Mg is readily leached from sandy and other soils having little cation exchange capacity. High levels of K or Ca in the soil also can induce Mg deficiency.

Occurrence: Mg deficiency is very common on highly leached soils in Florida and Hawaii. It can also occur in container-grown palms if dolomitic lime has not been added to the medium.

Species Affected: Most species of palms are affected, although *Syagrus* spp. and *Washingtonia* spp. seem to be highly resistant and *Phoenix* spp. particularly susceptible to Mg deficiency.

Diagnostic Techniques: Visual symptoms alone are usually sufficient for diagnosis except in the case of *Phoenix*, where Mg deficiency symptoms are very similar to those of K deficiency. See Appendices 10-12 for critical Mg concentrations.

Prevention and Treatment: Mg deficiency is difficult to correct once symptoms are present. It is best prevented by amending all container media with dolomite and by including Mg, preferably in a controlled-release form, in all palm fertilizers used in landscapes in regions prone to Mg deficiency. Treatment of deficient palms requires up to a year or more and is accomplished by applying magnesium sulfate at 1-2 kg per tree 4 to 6 times per year. As with K deficiency, symptomatic leaves will never recover and must be replaced by healthy new leaves. Foliar Mg sprays are generally ineffective in treating Mg deficiency since they supply very small amounts of Mg relative to the amount required by palms.

MANGANESE DEFICIENCY

Symptomology: The newest leaves of manganese (Mn) deficient palms emerge chlorotic with interveinal necrotic streaks. As the deficiency progresses, newly emerging leaflets appear necrotic and withered on all but basal portions of the leaflets. This withering results in a curling of the leaflets about the rachis giving the leaf a frizzled appearance ("frizzletop"). On new leaves of Mn-deficient *Cocos nucifera*, necrotic leaflet tips fall off and the leaf has a singed appearance. In severely Mn-deficient palms, growth stops and newly emerging leaves consist solely of necrotic petiole stubs. Meristem death usually follows.

Cause: Mn deficiency is caused by insufficient Mn in the soil or high pH, which reduces Mn availability. In soils where Mn is marginally sufficient, cold soil temperatures may cause temporary Mn deficiency by reducing root activity level relative to that of the foliage.

Occurrence: Mn deficiency is very common on alkaline soils, but can occur in containers if drainage is poor or soil temperatures are cool.

Species Affected: Most species of palms can be affected, but *Cocos nucifera*, *Syagrus* spp., *Roystonea* spp., *Acoelorrhaphe wrightii*, *Phoenix roebelenii*, *Elaeis guineensis*, and *Latania* spp. are particularly susceptible, while *Washingtonia* spp. appear resistant.

Diagnostic Techniques: Visual symptoms may be sufficient to diagnose this disorder, but leaf nutrient analysis is also suggested, since symptoms of B, Cu, and Zn deficiencies are similar. In *Syagrus* and *Roystonea* late stage K deficiency appears very similar to that of Mn deficiency. See Appendices 10-12 for critical Mn concentrations in some palms.

Prevention and Treatment: Fertilizers containing 0.5 to 1.0% Mn should be used routinely on soils where Mn deficiency is a problem. Although soil applications of manganese sulfate are effective, spraying the foliage may achieve more rapid short-term results.

NITROGEN DEFICIENCY

Symptomology: Early symptoms of nitrogen (N) deficiency include general loss of dark green foliage color and reduced growth rate. In severe cases, leaves will be almost completely yellow or whitish, growth will cease, and the palm will appear to be in general decline.

Cause: N deficiency is usually caused by insufficient N in the soil due to inadequate fertilization or excessive leaching.

Occurrence: N deficiency is common in both container-grown palms and in those growing in the ground.

Species Affected: Most palm species are susceptible to nitrogen deficiency.

Diagnostic Techniques: Visual symptoms and/or leaf nutrient analysis can be used to diagnose this disorder. See Appendices 10-12 for critical N concentrations for some palm species.

Prevention and Treatment: Nitrogen deficiency is best prevented by supplying ammonium, nitrate, or urea-containing fertilizers on a regular basis. Deficient palms can be treated with soil or foliar applications of N fertilizers. Unless frequent liquid fertilization is practiced, most N in fertilizers should be in a water-insoluble form to prevent rapid loss from leaching. Unlike most other nutrient deficiencies, N-deficient leaves usually regain their normal green coloration following treatment with N fertilizers.

PHOSPHORUS DEFICIENCY

Symptomology: Primary symptoms include a reduction in growth rate and gradual loss of green color. Severely affected palms show no signs of growth and in *Chamaedorea elegans* and *Elaeis guineensis*, purplish-brown necrotic spotting may appear on oldest leaves.

Cause: Phosphorus (P) deficiency is caused by insufficient available P in the soil. Soluble P fertilizers are rapidly leached from container media and may be precipitated by metallic cations in field

New leaflets of Mn-deficient *Archontophoenix alexandrae*.

New foliage of Mn-deficient *Rhapis excelsa*.

N-deficient *Chamaedorea elegans* (right).

N-deficient *Caryota mitis* (left).

N-deficient *Phoenix roebelinii* (left).

N-deficient *Chrysalidocarpus lutescens* (left).

P-deficient *Howea forsterana* (left).

P-deficient *Chrysalidocarpus lutescens* (left).

P-deficient *Caryota mitis* (left).

P-deficient *Chamaedorea elegans* (right).

soils. Insoluble P sources are only slightly available to plants.

Occurrence: P deficiency is rare in palms. P is generally adequate for palm growth in most soils in the United States and P is a major component of most mixed fertilizers used.

Species Affected: Most palm species can be affected.

Diagnostic Techniques: Visual symptoms plus leaf nutrient analysis are usually needed to diagnose P deficiency. Diagnosis by visual symptoms alone is difficult since symptoms are very similar to those of N deficiency. See Appendices 10-12 for critical P concentrations in some palm species.

Prevention and Treatment: Applications of P-containing fertilizers will prevent P deficiencies and will treat palms already deficient in P.

POTASSIUM DEFICIENCY

Symptomology: Symptoms of potassium (K) deficiency vary among species but always appear first on oldest leaves. Leaflets of some palms such as *Dictyosperma album* are mottled with yellowish spots that are translucent when viewed from below. In other palms such as *Arenga* spp. and *Roystonea* spp., symptoms appear on older leaves as a marginal or tip necrosis of the leaflets with little or no yellowish spotting present. The leaflets in *Roystonea* and other species showing tip necrosis often appear withered and frizzled.

In *Livistona chinensis* and *Bismarckia nobilis*, necrosis is not marginal, but is confined to the centers and tips of the leaflets. In *Phoenix* spp., the distal parts of the oldest leaves are typically orange with leaflet tips becoming necrotic. The rachis of the leaves remains green, however, and the orange and green are not sharply delimited as with Mg deficiency.

In *Caryota mitis*, chlorotic mottling is minimal or nonexistent, but early symptoms appear as necrotic streaks within the leaflets. In most other palm species, including *Cocos nucifera*, *Elaeis guineensis*, *Neodypsis decaryi*, *Chamaerops humilis*, *Chrysalidocarpus* spp., *Livistona mariae*, and *Hyophorbe verschafeltii*, early symptoms appear as translucent yellow or orange mottling of the leaflets and may be accompanied by necrotic spotting. In more severely deficient palms, marginal or marginal and tip necrosis will be present. The most severely affected leaves or leaflets will be completely necrotic and withered in appearance.

Cause: K deficiency is caused by insufficient K in the soil, but can be induced by high N, Ca, or Mg to K ratios in the soil.

Occurrence: K deficiency is very common on palms growing in highly leached sandy soils or in container media. It is relatively uncommon on palms growing in clay. K is retained against leaching in soils having moderate cation exchange capacity, but is readily leached from sands or soils having little cation exchange capacity.

Species Affected: Most palms are quite susceptible to K deficiency. More tolerant palms include *Archontophoenix alexandrae* *Trithrinax acanthocoma*, and *Gaussia maya*.

Diagnostic Techniques: Visual symptoms alone are usually sufficient for diagnosis, although leaf nutrient analysis may be helpful

Older leaf of K-deficient *Chrysalidocarpus lutescens*.

Older leaf of K-deficient *Chamaedroea elegans*.

Older leaf of K-deficient *Dictyosperma album*.

Older leaflets of K-deficient *Arenga* sp.

Older leaflets of K-deficient *Caryota mitis*.

Older leaf of K-deficient *Cocos nucifera*.

K-deficient *Roystonea regia* (right).

K-deficient *Hyophorbe verschafeltii*.

Powerline decline of *Roystonea regia*.

Powerline decline of *Cocos nucifera*.

Dying *Phoenix canariensis* due to poor drainage following transpolanting.

Shrivelled trunk of *Syagrus romanzoffiana* planted too deeply.

Cocos nucifera growing in poorly drained soil.

Paurotis sp. planted too deeply.

in the case of *Phoenix* where K deficiency resembles Mg deficiency, or in *Caryota* where symptoms are similar to those of some leafspot diseases. See Appendices 10-12 for critical K concentrations in some palm species.

Prevention and Treatment: Regular applications of K fertilizers will prevent K deficiency and treat palms already deficient in K. On sandy soils, or those having little cation exchange capacity, K fertilizer should be applied in coated controlled-release forms to prevent rapid loss due to leaching. Fertilizers having a 3N-1P-3K-1Mg ratio appear to be best for palms.

POWERLINE DECLINE

Symptomology: Leaves near high voltage powerlines are chlorotic with necrotic tips in the vicinity of the lines. In severe cases the entire crown will become chlorotic and may die.

Cause: Electro-magnetic fields around high voltage power lines appear to injure palms. Foliage within 1-2 feet of a line may be injured. Actual contact with the wires is not necessary for injury to occur.

Occurrence: Powerline decline is common wherever palms are planted under powerlines.

Species Affected: Most species can be affected, but only taller species will likely encounter overhead powerlines.

Diagnostic Techniques: Visual symptoms are sufficient for diagnosis.

Prevention and Treatment: This problem can be prevented by planting palms far enough from overhead wires that the foliage will never be able to contact the lines. If a palm is directly under a powerline, it should be removed, but if sufficiently far away that only leaf tips are affected, the palm may eventually grow above the lines and recover from the symptoms.

ROOT SUFFOCATION

Symptomology: Early symptoms of root suffocation are often those of Fe deficiency with chlorotic new leaves emerging. In severe cases, wilting of the foliage and shrinkage of the trunk may occur. Symptoms may not appear until years after planting.

Cause: Planting palms more deeply than they were originally growing decreases aeration in the root zone and causes root suffocation. This can lead to eventual death of the root system and of the entire palm. Waterlogged soils reduce soil aeration with similar effects on palm roots.

Occurrence: Palms in poorly drained landscape or field sites, or in poorly drained container media often have this problem.

Species Affected: Most species of palms are susceptible to root suffocation.

Diagnostic Techniques: Visual symptoms are usually adequate for diagnosis of this problem. Look for Fe deficiency symptoms, wilting, shriveling of the trunk, or excessive water in the root zone. Dig down to determine if the palm was planted too deeply.

Prevention and Treatment: Dig out deeply planted palms and replant at the proper depth. Poorly drained sites should be built up with a berm before planting.

SULFUR DEFICIENCY

Symptomology: New leaves emerge uniformly yellow and slightly reduced in size. As the deficiency progresses, leaflet tips become necrotic and growth is stunted. Older leaves remain green in all but the most severely deficient palms. Symptoms are similar to those of N and Fe deficiency.

Cause: Sulfur (S) deficiency is caused by insufficient S in the soil. S in sulfate form is readily leached from all soils.

Occurrence: S deficiency is relatively uncommon, but can occur in highly leached soils or container media if no sulfate fertilizers are used.

Species Affected: S deficiency can occur in any species.

Diagnostic Techniques: Visual symptoms may be confused with N or Fe deficiency symptoms, so leaf nutrient analysis is helpful in diagnosing this disorder. See Appendices 10-12 for critical S concentrations in some palm species.

Prevention and Treatment: S deficiency is easily prevented and treated by using sulfate-containing fertilizers on a regular basis. Fertilizers should contain about as much S as P to prevent S deficiency. Mildly deficient foliage may completely recover following treatment, but any necrotic tissue will remain.

SUNBURN

Symptomology: Large necrotic areas are visible on the upper surfaces of leaves, usually in the center of leaves or leaflets, rather than on leaf tips or margins. Affected foliage on adjacent leaves will often have the same directional orientation.

Cause: Exposure of shade-grown foliage to high light intensities can injure leaves.

Occurrence: Sunburn is fairly common on palms moved from shade to full sun growing conditions.

Species Affected: Sunburn can affect most species of palms.

Diagnostic Techniques: Visual symptoms alone are usually sufficient to diagnose this disorder.

Prevention and Treatment: Sunburn can be prevented by growing palms in full sun if they are to be used in the landscape. Acclimatization of palm leaves involves replacement of the entire canopy by leaves adapted to the new higher light intensity. Individual shade-grown leaves cannot adapt to higher light intensities.

New leaves of S-deficient *Syagrus romanzoffiana*.

S-deficient *Caryota mitis* (left).

S-deficient *Chamaedorea elegans* (left).

S-deficient *Howea forsterana*.

Exposed leaf of sunburned *Chamaedorea elegans*.

Geonoma sp. suffering water stress.

Zn-deficient *Chrysalidocarpus lutescens*.

New leaves of Zn-deficient *Chamaedorea elgans*.

Zn-deficient *Howea forsterana*.

Zn-deficient *Phoenix roebelinii*.

WATER STRESS

Symptomology: Typical water stress symptoms include reduced growth and necrosis on leaflet tips, spreading to the entire leaf as severity increases. Newly emerging leaves may wither and die. Death of the meristem may follow. Water stress in some species is first indicated by leaflets folding or wilting.

Cause: Water stress occurs when water is limited or the root system is incapable of taking up sufficient water.

Occurrence: Water stress is fairly common in container-grown palms, but is less common in landscape or field situations. Palms in general are fairly drought tolerant once they become established.

Species Affected: Water stress is generally most severe on species such as *Geonoma* spp., that are native to tropical rain forests. However, many *Chamaedorea* spp. are extremely drought tolerant.

Diagnostic Techniques: Visual symptoms are usually sufficient to diagnose this problem.

Prevention and Treatment: Water stress is easily prevented by not allowing the soil to become excessively dry. Mildly stressed palms will usually recover when watered and only lose some foliage, but severely stressed palms of sensitive species may not recover.

ZINC DEFICIENCY

Symptomology: New leaves show interveinal chlorosis in zinc (Zn) deficient *Chamaedorea elegans* and *Phoenix roebelenii*, but not in *Chrysalidocarpus lutescens* or *Howea forsterana*. In more severely deficient palms, leaflet tips become necrotic, the necrosis increasing until only necrotic petiole stubs remain. Death of the meristem will occur if corrective treatment is not given.

Cause: Zn deficiency is caused by insufficient available Zn in the soil. High soil pH can precipitate Zn and render it unavailable to palms.

Occurrence: Zn deficiency is rare in landscapes in the United States, but can occur in container-grown palms if no Zn fertilizers have been used.

Species Affected: Most species of palms are probably susceptible to Zn deficiency.

Diagnostic Techniques: Leaf nutrient analysis is usually needed in addition to visual symptoms for accurate diagnosis of Zn deficiency. See Appendices 10-12 for critical Zn concentrations in some palms.

Prevention and Treatment: Zn deficiency can be prevented by using fertilizers containing Zn. Treatment of deficient palms with zinc sulfate applied to the soil or foliage is usually effective.

Appendix 1. Chu medium No. 10 for culturing *Cephaleuros*.

Ingredient	Amount g/L	Ingredient	Amount g/L
Ca(NO$_3$)$_2$	0.004 g/L	K$_2$HPO$_4$	0.01 g/L
MgSO$_4$:7H$_2$O	0.025 g/L	Na$_2$CO$_3$	0.02 g/L
Na$_2$SiO$_3$	0.025 g/L	FeCl$_3$	0.0008 g/L

A modification of ferric citrate (0.003 g/L) and citric acid (0.003 g/L) is suggested for the ferric chloride component.

Appendix 2. Palm taxa reported as hosts of *Bipolaris* spp.

Bipolaris cynodontis	*Bipolaris incurvata*	*Bipolaris setariae*
Archontophoenix alexandrae	*Archontophoenix alexandrae*	*Caryota mitis*
Archontophoenix cunninghamiana	*Archontophoenix cunninghamiana*	*Chamaedorea elegans*
Caryota mitis	*Caryota mitis*	*Chrysalidocarpus lutescens*
Chamaedorea elegans	*Chamaedorea seifrizii*	*Phoenix roebelenii*
Chrysalidocarpus lutescens	*Chrysalidocarpus lutescens*	
Howea forsterana	*Cocos nucifera*	
Rhapis sp.	*Howea belmoreana*	
	Howea forsterana	
	Livistona chinensis	
	Licuala ramsayi	
	Phoenix roebelenii	
	Ptychosperma elegans	
	Rhapis sp.	
	Roystonea regia	
	Syagrus romanzoffiana	
	Washingtonia robusta	

Appendix 3. Palm taxa reported as hosts of *Exserohilum rostratum* or *Phaeotrichoconis crotalariae*.

Exserohilum rostratum	*Phaeotrichoconis crotalariae*
Archontophoenix alexandrae	*Caryota mitis*
Archontophoenix cunninghamiana	*Chamaedorea elegans*
Caryota mitis	*Chrysalidocarpus lutescens*
Chamaedorea elegans	*Livistona chinensis*
Chamaedorea seifrizii	
Chrysalidocarpus lutescens	
Elaeis guineensis	
Howea forsterana	
Ptychosperma elegans	
Rhapis excelsa	
Roystonea regia	
Syagrus romanzoffiana	
Washingtonia robusta	

Appendix 4. Palm taxa known to be hosts of *Ganoderma zonatum*, *Graphiola phoenicis*, or *Pestalotiopsis palmara*.

Ganoderma zonatum	*Graphiola phoenicis*	*Pestalotiopsis palmara*
Acoelorrhaphe wrightii	*Acoelorrhaphe wrightii*	*Bismarckia nobilis*
Acrocomia aculeata	*Arenga pinnata*	*Butia capitata*
Areca catechu	*Butia capitata*	*Caryota mitis*
Attalea sp.	*Chamaerops humilis*	*Caryota urens*
Bactris sp.	*Chrysalidocarpus lutescens*	*Chamaedorea elegans*
Borassus aethiopium	*Cocos nucifera*	*Chrysalidocarpus lutescens*
Brahea berlandieri	*Coccothrinax argentata*	*Cocos nucifera*
Butia capitata	*Phoenix canariensis*	*Elaeis guineensis*
Butia yatay	*Phoenix dactylifera*	*Phoenix canariensis*
Chamaerops humilis	*Phoenix reclinata*	*Phoenix dactylifera*
Chysalidocarpus cabadae	*Phoenix roebelenii*	*Phoenix reclinata*
Chrysalidocarpus lutescens	*Phoenix sylvestris*	*Phoenix roebelenii*
Cocos nucifera	*Phoenix theophrasti*	*Rhapis excelsa*
Copernicia curtisii	*Roystonea elata*	*Roystonea elata*
Gastrococos crispa	*Sabal minor*	*Sabal palmetto*
Livistona chinensis	*Sabal palmetto*	*Veitchia merrillii*
Phoenix canariensis	*Syagrus romanzoffiana*	*Washingtonia* spp.
Phoenix reclinata	*Washingtonia robusta*	
Phoenix sylvestris		
Polyandrococos caudescens		
Ptychosperma elegans		
Ptychosperma lineare		
Ptychosperma macarthurii		
Ptychosperma salomonense		
Roystonea altissima		
Roystonea elata		
Roystonea regia		
Sabal causiarum		
Sabal palmetto		
Scheelea sp.		
Serenoa repens		
Syagrus oleracea		
Syagrus romanzoffiana		
Syagrus romanzoffiana var. *australe*		
Syagrus schizophylla		
Washingtonia robusta		

Appendix 5. Susceptibility of palm taxa to Lethal Yellowing Disease.

Known to be susceptible	Not known to be susceptible
Aiphanes lindeniana	*Acoelorrhaphe wrightii*
Allagoptera arenaria	*Arenga pinnata*
Arenga engleri	*Bactris gasipaes*
Borassus flabellifer	*Carpentaria acuminata*
Caryota mitis	*Chamaerops humilis*
Chrysalidocarpus cabadae	*Chrysalidocarpus lutescens*
Cocos nucifera	*Coccothrinax argentea*
Corypha elata	*Elaeis guineensis*
Corypha taliera	*Heterospathe elata*
Dictyosperma album	*Phoenix roebelenii*
Gaussia attenuata	*Ptychosperma macarthurii*
Howea belmoreana	*Rhapidophyllum hystrix*
Hyophorbe verschaffeltii	*Roystonea elata*
Latania sp.	*Roystonea hispaniola*
Livistona chinensis	*Roystonea regia*
Livistona rotundifolia	*Sabal causiarum*
Nannorrhops ritchiana	*Sabal palmetto*
Neodypsis decaryi	*Syagrus amara*
Phoenix canariensis	*Syagrus romanzoffiana*
Phoenix dactylifera	*Thrinax morrisii*
Phoenix reclinata	*Thrinax parviflora*
Phoenix sylvestris	*Washingtonia filifera*
Pritchardia affinis	*Washingtonia robusta*
Pritchardia pacifica	
Pritchardia remota	
Ravenea hildebrandtii	
Syagrus schizophylla	
Trachycarpus fortunei	
Veitchia arecina	
Veitchia merrillii	
Veitchia montgomeryana	

Appendix 6. Antibiotic injection methods for control of Lethal Yellowing Disease.

Gravity-Flow injection--antibiotic is dissolved in 8-16 fl oz of water in a solution reservoir like an empty bleach container. The entire solution flows by gravity through plastic tubing into the tree overnight. This method is applicable to urban settings with few trees. Injection can be erratic.

Mauget injection--commercially available, single use plastic containers attached to a feeder tube that is inserted into a pre-drilled hole in the palm trunk. This process can inject 2 fl oz of active antibiotic in a 4-12 hour period. The injection capsule should be left on the palm for a week before removal and disposal.

Minute-Tree injection--commercially available, reusable, modified grease gun or hydraulic jack. This injector can be loaded with 1 oz of antibiotic in 0.5-1.0 fl oz of water. The needle is inserted into a pre-drilled hole in the trunk and the handle is pumped to force the solution into the trunk in about one minute. The process is repeated to deliver the full dosage. This is a rapid, labor-efficient method for treating large numbers of trees. High pressures generated by this method can cause internal tissue damage and trunk splitting. Do not reuse these injector sites.

Self-Contained, Air-Pressure injection--relies on a solution reservoir of 1 qt or more made from either a compressed air cylinder, fire extinguisher, or pipe section that can hold a pressure of 100 p.s.i. The reservoir is connected to a hollow lag screw by a quick connect coupler and can be pressurized by hand pump or compressed air source. The injection requires approximately 30-60 minutes.

Appendix 7. Palm taxa reported as hosts of *Phytophthora palmivora*.

Archontophoenix alexandrae	*Livistona rotundifolia*
Borassus flabellifer	*Normanbya normanbyi*
Butia capitata	*Phoenix canariensis*
Chamaedorea elegans	*Pinanga insignis*
Chamaedorea erumpens	*Ptychosperma macarthurii*
Chamaedorea seifrizii	*Rhopalostylis* sp.
Chrysalidocarpus lutescens	*Roystonea* sp.
Cocos nucifera	*Sabal* sp.
Elaeis guineensis	*Syagrus romanzoffiana*
Howea forsterana	*Trachycarpus fortunei*
Kentia sp.	*Washingtonia filifera*
	Washingtonia robusta

Appendix 8. Some ammendments for corn meal agar used to isolate *Phytophthora* spp.

Medium		Rate per liter
1	pimaricin	10 ppm
	penicillin	50 ppm
	pentachloronitrobenzene (PCNB)	100 ppm
2	pimaricin (90% a.i.)	10 mg
	vancomycin (100% a.i.)	200 mg
	PCNB (75% a.i.)	100 mg
	hymexazol (98-99% a.i.)	50 ppm
3	pimaricin (50% a.i.)	10 mg
	ampicillin (81% a.i.)	250 mg
	rifampicin (100% a.i.)	10 mg
	hymexazol (98-99% a.i.)	50 mg
	PCNB (75% a.i.)	100 mg

Media 2 and **3** are more selective than **Medium 1** for *Phytophthora* isolation. **Medium 3** is more economical than **Medium 2**. The addition of hymexazol (at 25-50 ppm) partially inhibits recovery of *Phytophthora palmivora*.

Appendix 9. Palm taxa known to be hosts of *Stigmina palmivora*.

Acoelorrhaphe wrightii
Borassus aethiopium
Butia capitata
Caryota mitis
Caryota urens
Chrysalidocarpus sp.
Cocos nucifera
Howea forsterana
Livistona sp.
Phoenix canariensis
Phoenix dactylifera
Phoenix loureirii
Phoenix reclinata
Phoenix roebelenii
Phoenix rupicola
Rhapis exselsa
Roystonea elata
Roystonea regia
Sabal palmetto
Syagrus romanoffiana
Thrinax morrisii
Veitchia merrillii
Washingtonia robusta

Appendix 10. Critical concentrations of 13 elements in *Chamaedorea elegans, C. erumpens* and *Chrysalidocarpus lutescens*. Concentrations for elements N through Na expressed in percent; those for Fe through Zn are given in ppm.

Element	Deficient	Low	Normal	High	Excessive
N	1.90	2.0-2.40	2.50-3.50	3.60-4.50	4.50 +
S	0.14	0.15-0.20	0.21-0.40	0.41-0.75	0.76 +
P	0.10	0.11-0.14	0.15-0.30	0.31-0.75	0.76 +
K	1.20	1.25-1.55	1.60-2.75	2.80-4.00	4.05 +
Mg	0.20	0.21-0.24	0.25-0.75	0.76-1.00	1.01 +
Ca	0.39	0.40-0.99	1.00-2.50	2.51-3.25	3.26 +
Na	--	--	0-0.20	0.21-0.50	0.51 +
Fe	39	40-49	50-300	301-1000	1001 +
Al	--	--	0-250	251-2000	2001 +
Mn	39	40-49	50-250	251-1000	1001 +
B	17	18-24	25-60	61-100	101 +
Cu	3	4-5	6-50	51-200	201 +
Zn	17	18-24	25-200	201-500	501 +

Appendix 11. Critical concentrations for 13 elements in *Howea forsterana* and *Rhapis excelsa*. Concentrations for elements N through Na are expressed in percent; those for Fe through Zn are given in ppm.

Element	Deficient	Low	Normal	High	Excessive
N	0.84	0.85-1.19	1.20-2.75	2.76-4.00	4.01 +
S	0.10	0.11-0.14	0.15-0.75	0.76-1.25	1.26 +
P	0.10	0.11-0.14	0.15-0.75	0.76-1.25	1.26 +
K	0.59	0.60-0.84	0.85-2.25	2.26-4.00	4.01 +
Mg	0.19	0.20-0.24	0.25-1.00	1.01-1.25	1.26 +
Ca	0.25	0.26-0.39	0.40-1.50	1.51-2.50	2.51 +
Na	--	--	0-0.20	0.21-0.50	0.51 +
Fe	39	40-49	50-250	251-1000	1001 +
Al	--	--	0-250	251-2000	2001 +
Mn	39	40-49	50-250	251-1000	1001 +
B	15	16-20	21-75	76-100	101 +
Cu	4	5-7	8-200	201-500	501 +
Zn	17	18-24	25-200	201-1000	1001 +

Appendix 12. Critical concentrations for 13 elements in *Elaeis guineensis*. Concentrations for elements N through Na are expressed in percent; those for Fe through Mo are given in ppm.

Element	Deficient	Low	Normal
N	1.90	2.00-2.30	2.40-3.00
P	0.11	0.12-0.14	0.15-0.25
K	0.74	0.75-0.99	1.00-1.80
Mg	0.19	0.20-0.23	0.24-0.35
Ca	0.29	0.30-0.59	0.60-1.00
Na	--	--	0-0.20
Fe	29	30-39	40-100
Al	--	--	0-250
Mn	99	100-149	150-300
B	4	5-7	8-30
Cu	2	3	4-20
Zn	9	10-13	14-100
Mo	--	--	1-10

Appendix 13. Common names of some palm taxa.

Acoelorrhaphe wrightii	Paurotis or Everglades Palm
Allagoptera arenaria	Seashore Palm
Archontophoenix alexandrae	Alexandra, King or Northern Bangalow Palm
Archontophoenix cunninghamiana	Piccabeen Palm
Areca catechu	Betel Nut Palm
Arenga pinnata	Sugar, Areng or Black-fiber Palm
Bactris gasipaes	Peach Palm
Borassus flabellifer	Palmyra Palm
Butia capitata	South American Jelly or Pindo Palm
Carpentaria acuminata	Carpentaria Palm
Caryota mitis	Burmese, Clustered, or Tufted Fishtail Palm
Chamaedorea elegans	Parlor Palm
Chamaedorea erumpens	Bamboo Palm
Chamaedorea seifrizii	Bamboo or Reed Palm
Chamaerops humilis	European Fan Palm
Chrysalidocarpus cabadae	Cabada Palm
Chrysalidocarpus lutescens	Areca, Butterfly or Golden Palm
Coccothrinax argentata	Silver Palm
Cocos nucifera	Coconut Palm
Corypha elata	Buri Palm
Dictyosperma album	Hurricane or Princess Palm
Elaeis guineensis	African Oil Palm
Gaussia attenuata	Puerto Rican Gaussia Palm
Heterospathe elata	Sagisi Palm
Howea belmoreana	Belmore Palm
Howea forsteriana	Forster Sentry or Kentia Palm
Hyophorbe lagenicaulis	Bottle Palm
Hyophorbe verschaffeltii	Spindle Palm
Latania sp.	Latan Palm
Livistonia chinensis	Chinese Fan Palm
Nannorrhops ritchiana	Mazari Palm
Neodypsis decaryi	Triangle Palm
Phoenix canariensis	Canary Island Date Palm
Phoenix dactylifera	Date Palm
Phoenix reclinata	Senegal Date Palm
Phoenix roebelenii	Dwarf, Miniature, or Pygmy Date Palm
Phoenix sylvestris	Wild Date Palm
Pritchardia affinis	Kona Palm
Pritchardia pacifica	Fiji Island Fan Palm
Ptychosperma elegans	Alexander or Solitaire Palm
Ptychosperma macarthurii	Macarthur Palm
Rhapidophyllum hystrix	Needle Palm
Rhapis excelsa	Lady Palm
Rhapis subtilis	Thai Dwarf Lady Palm
Roystonea elata	Florida Royal Palm
Roystonea hispaniola	Hispaniolan Royal Palm
Roystonea regia	Cuban Royal Palm
Sabal causiarum	Puerto Rican Hat Palm
Sabal palmetto	Cabbage Palmetto or Florida Cabbage Palm
Serenoa repens	Saw or Scrub Palmetto
Syagrus romanzoffiana	Queen Palm
Syagrus schizophylla	Arikury Palm
Thrinax morrisii	Key Thatch Palm
Thrinax parviflora	Jamaica Thatch Palm

Thrinax radiata	Florida Thatch Palm
Trachycarpus fortunei	Chinese Windmill, Hemp or Windmill Palm
Veitchia merrillii	Christmas, Merrill or Manila Palm
Veitchia montgomeryana	Montgomery's Palm
Washingtonia filifera	California Fan Palm
Washingtonia robusta	Mexican Fan Palm